I0064009

NEW HOPE
FOR
CONCUSSIONS TBI & PTSD

BY
DR. LAWRENCE D. KOMER
JOAN CHANDLER KOMER

First Edition

This book provides general information. Always consult your health professionals for advice on what is best for you.

Published By
Peak Performance Institute Inc.
www.peak-performance-institute.com
Toronto

ISBN 978-0-9952501-2-3

Cover design by Todd Reny

Cover Photo digitalista/bigstock.com

Printed in the United States of America

Dedication

This book is dedicated to our children Scott Alexander Komer and Kimberly Dawn Komer. You both have made a difference to the world, and through our book will continue to help others. We love you both so very much.

Scott and Kim Komer

Kim Komer

Scott Komer

CONTENTS

TABLE OF CONTENTS

FOREWORD

I met Larry and Joan Komer in the early 1990s, back in the days before the first Chicken Soup for the Soul® book was published. From the beginning of our relationship, it was obvious the Komers cared about making a difference in the life of others.

Together, they gave our first seminar for physicians and their spouses called, "Chicken Soup for the Doctor's Soul." They went on to do many educational seminars, including their popular, "Be Menopositive!" and "Be Andropositive!" Their seminars are full of information and inspiration around health and wellness empowering and caring for the thousands of souls who have attended over the years.

As a physician, Larry was always curious about what medicine could do to help his patients be well and thrive. His curiosity and insights made him a leader in both men and women's health, and now in the area of brain health as well. Larry's genuine care for his patients as people was the impetus behind his well-informed discoveries and innovations. That is a rare trait in today's world of healthcare.

Joan, a long-time educator with a special interest in self-esteem, has impacted the lives of so many through motivating, teaching, encouraging and comforting others. She is just one of those people who make you feel better.

The work that the Dr. Larry and Joan Komer are doing in brain health addresses an epidemic of trauma, and pain inflicted on individuals and their families worldwide. Their insights and lessons are meant to encourage the heart and souls of those on their journey to wellness.

"New Hope for Concussions TBI & PTSD" is a powerful resource for the injured, the caregivers, the sporting world, the medical community, and those serving our veterans and others with PTSD. It is a book of hope for all those who have been told, "We are sorry but there is nothing more we can do."

My hope for anyone reading this book is to never give up and continue to strive for the best!

Jack Canfield
Co-Creator, Chicken Soup for the Soul Book Series

ACKNOWLEDGEMENTS

Many people were part of this book becoming a reality. We would like to thank our friend, Jack Canfield, for his kind and generous Foreword in the book. Our colleagues, Dr. William Cook, Dr. Blair Lamb, Andrew Marr, and Dr. Patrick Quaid, each contributed a chapter to make this book a better resource to understand the brain.

The many thousands of patients over the years are part of our extended family. We learned from you as we served you. It has been a privilege. Any specific patient references were used with the patient's permission.

Grant D. Fairley of McK Consulting Inc. served as our writing coach and project manager for this book. Over the years, Grant has been a great friend and inspiration. Without his guidance, this book would not have been a reality.

We would especially like to thank our senior editor, Jenny Fountain, MA, for her excellent work. Our thanks also go to the other line editors and reviewers. Numerous other people were kind enough to read the early drafts of the book and gave us helpful suggestions.

Our son, Scott Komer, managed our cover design process and coordinated our social media. He has been a valuable advisor in many aspects of this book. The cover design was done by Todd Reny, with the cover image from digitalista/bigstock.com

Our office team at The Komer Clinics have been indispensable over the years and especially as we added this project to our busy practice. Our special thank-you goes to Heather Lawson, Sarah Hepworth and Gord Tonnelly.

I also want to acknowledge over three decades of friendship and unfailing support from many of my physician colleagues, especially Drs. Ben Carruthers, Randy Cross, Tom Durcan, Jim Faught, Barry Hunter, David Levy, David Sansome, and Frank Stechey.

Heather Lawson **Gord Tonnelly** **Sarah Hepworth**

(Images above by Michael Dismatsek www.dmnikas.ca)

INTRODUCTION

I used to think that I was easily bored. There was always an urge for something new and different. It seemed that every five years, I wanted to take on an interesting challenge. There were many choices because of the range of research and practice in Obstetrics and Gynecology. Looking back, I now understand that what I had been given was the gift of curiosity from my father, Stan. That curiosity would take me on an amazing adventure in healthcare, and open me to wide interest in many areas beyond medicine.

As someone who has always loved sports as a player, coach, and later as a team physician, I became increasingly concerned with the growing incidence of concussions. They were having an impact on player health. Dealing with patients of all ages in my women's and men's practice, I regularly met people who had suffered brain traumas.

While attending conventions on men's health, I was fortunate to meet Dr. John Crisler of the "All Things Male-Center for Man's Health" in Lansing, Michigan. We hit it off right away. Together, we explored ideas and our treatment protocols over many years. He is a leading practical practitioner of men's health in the United States. Beyond his insights as a physician, John is one of those people who energize you every time you talk together. His enthusiasm for maximizing men's health is an inspiration for others.

As coach and physician of numerous hockey and lacrosse teams, I was concerned when my son Scott wanted to play hockey. His interest was not just playing hockey but also to be a goalie. From personal experience,

I knew the difficulties of playing goal in both lacrosse and hockey. I tried to divert him away from it. However, Scotty was committed to playing hockey, and so he joined a team. My next tactic was to encourage him to be a forward, but he saw my goalie equipment at home. (This equipment I refer to as "the tools of ignorance!") He tried out for the position and made it. By age 5, he also played lacrosse on a team with his sister, Kim. Before long, he became a lacrosse goalie too.

Like most active children, he encountered his share of challenges. He suffered a serious setback when he broke his femur skiing at age ten. Scotty would spend 11 weeks in traction followed by ten weeks in a body cast. At that point, he was in a very competitive hockey program. When someone is injured and out of hockey for a whole season at this important stage in development, it becomes very unlikely that he would be able to make a high level team the next season. So his career went from city club teams to house league teams. However, Scotty continued to improve. By the beginning of high school, he surprised everyone when he made the senior team. This hockey career has taken him to the present day where he is playing with retired NHL players.

As a lacrosse goalie, Scotty excelled from the beginning. He became a star in our hometown of Burlington, Ontario in Minor and later Junior A divisions. He was drafted by the Buffalo Bandits Pro Lacrosse Team and made the team. In the summer lacrosse league, he won a Canadian Championship. Over the years, he played some seasons in Europe and won a championship there. Another highlight was the opportunity to play in the World Indoor Lacrosse Championships in 2015.

My involvement in lacrosse included being the Team Physician on Scotty's minor lacrosse programs teams, Junior A teams and tournament teams in Prague and the Men's Senior B Canadian champions from St. Catharines in 2013. I looked after the Major Senior Lacrosse Team in Brampton where Scott coached. It was a privilege to be the Team Physician for the

Toronto Rock Professional Lacrosse Team for 17 years that included five championships. For Team Canada, I cared for the teams of the Men's Indoor Lacrosse gold medal champions in the 2003 and 2015 and the Team Canada Women's Outdoor Lacrosse silver medal champions in 2013.

These many experiences prompted me to start the Komer Brain Science Institute in 2015 to research and treat brain injuries. I know first-hand what it is like to have concussions playing hockey. My son has had concussions in both lacrosse and hockey as have numerous players in both sports that I treated. My daughter Kim had a serious concussion when her car was hit by a drunk driver. I saw the impact that these traumas could have on people of all ages, and especially young lives.

I came to this treatment from a unique perspective.

Some neurologists said that there is no treatment for concussions.

I disagree completely.

You will find out in this book why. I was motivated to discover a solution. My protocol with brain-injured patients and athletes has shown remarkably good results.

Joan shares insights about how to cope with the stresses of an injury. Concussions, TBIs, and PTSD do not just impact the individual. It can change families and whole communities. These chapters will help you, and those you love, to understand some of the personal dynamics of managing an injury, as you journey to wellness. The principles that she highlights will help you find greater peace as you recover.

The chapters by our esteemed friends and colleagues will also enable you to learn about different elements of brain health. We are pleased to include their important insights and experience from their practices. Andrew Marr's chapter on being a wounded warrior will illuminate these battlefield injuries and their consequences in a very dramatic way.

A CRACK IN TIME

There are moments in time when the course of history is changed. The Titanic struck an iceberg. Pearl Harbor was attacked. The atomic bomb was dropped. The assassinations of President Lincoln, Kennedy, Martin Luther King and Robert Kennedy traumatized a nation with each tragedy. Humans landed on the moon on. The Berlin Wall came down. 9-11 became more than a date. The day before, and the days after these events never were the same. These moments were a crack in time.

How dramatically life was about to change was not clear at the beginning of these events. Only later, did we begin to comprehend how much an instant may trigger a whole new destiny.

As a crack in time occurs in history, it happens to us as individuals too. Some of these can be very happy changes. We graduate. A new job begins. We marry. A friendship is made. We welcome a new life into the family. There are so many moments that can signal a better future.

Painful events may also create a "life before" and a "life after" moment. We say good-bye to a grandparent, parent, spouse, and most tragically, a child. We lose our job. Our retirement date is reached. Life goes on, but nothing will ever be quite the same. It is our personal crack in time.

There are moments when our brain health changes too. There was the way we were before the car accident and the way we were after it. Our life before the major concussion in the football game is transformed by the clash of helmets. The concussive explosion created a different world for you after that battlefield injury.

We try to remember what it used to be like. The many "if onlys" loop in our thoughts to imagine what would have been different had this horrible moment not been part of our story. At times, it is like trying to remember the face of a long-departed loved one, or the sound of their voice. How can it be this difficult?

While many life events create a clear line between before and after, some significant moments are not that obvious. The car's tire runs over a nail on the highway. It may be a while before you find out the tire has gone flat, and you are stranded. When the tropical depression that would become Hurricane Katrina formed over The Bahamas on August 23, 2005, no one in New Orleans could begin to imagine how their area would be devastated by the coming storm and the flooding that followed.

Some health events are easy to pinpoint, while others are more subtle and go unrecognized for a long time. This is especially true with concussions and other brain injuries that occur in sports, accidents, in a fall, or on the battlefield. We may not realize the damage that was done to our brain at the time of the incident until much later. Concussions, and other brain injuries, have their own memory – even the seemingly small ones add up. When enough damage is done, the symptoms of concussion become obvious. We can then look back and see how all these minor moments have accumulated into our major health crisis.

Whether it is a dramatic injury, or the accumulation of small knocks to your head, at some point you come to realize that life has changed for you. Your future has new limitations. Your everyday routines are colored by brand new aches and pains. Even your mood might change. It will never again be the same.

It is your crack in time.

BAD NEWS

You slipped, fell and hit your head. Now, you have a bit of a headache, but you were not dazed or unconscious. A bit tired, you just feel "off" for a couple of days. Over time, it seems to get better. Within a few days or weeks, you feel right back to normal. However, the reality is that you may have had a concussion. There is no blood test to diagnose concussion. There are questionnaires such as ImPACT or SCAT3 you can use, and they might suggest a concussion. You may be surprised to learn that often one is not instantly sure.

Perhaps, you were involved in a motor vehicle accident. Shaken up and sore, you thank Heaven nothing more serious happened. There is that bit of a sore neck, but this mild whiplash can go away.

Were you someone involved in sports? Did you collide with someone? Were you checked into the boards playing hockey or lacrosse? You missed a few shifts. Then you were back in the game. It would not be unusual if you do not remember very much of what happened, or the rest of the day.

Was soccer (or football as it is known in much of the world) your game? You never had any problems during practice or the game heading the ball. Over many years, this was part of your game. No harm done. Or was there?

As a young child, did you fall down the stairs and hit your head?

There could be bad news associated with each of these events.

Every one of them could have been a concussion.

It is not difficult to diagnose a concussion when a serious event has taken place. If you hurt your head, you might have problems in two or three weeks, and have been lingering effects such as headache, or head pressure or fatigue, or difficulties concentrating, the diagnosis of concussion is fairly certain.

It turns out that repeated mild blows to the head injuries can add up to long-term difficulties, triggering a change in brain function. The more severe the injury is the greater risk of developing serious problems.

When I speak about sports concussions, people think about boxing, football, hockey, lacrosse or motocross. However, do you know that one of the most dangerous sports is cheerleading? Concussions are frequent in many sports, like gymnastics.

A helmet does not save you from a concussion. It may save a skull fracture or a laceration but when your head accelerates then suddenly stops, your brain hits against the inside of the skull. That can cause a concussion.

This all connects to what you hear on the news about the dangers of post-concussion syndrome. There is a growing understanding that this may also create an increased risk of things like Parkinson's disease and chronic traumatic encephalopathy (CTE).

First responders and those serving in the military have double the risk. People in public safety roles are often in situations where they experience head injuries. They are placed in dangerous situations where they may be struck with objects, or assaulted by people they are trying to apprehend. Traveling at high speeds in an emergency, to help or protect others, they may crash their vehicles. Military personnel may also be hit by objects or be subject to blast injuries. When a bomb goes off, there is a concussive wave that can cause multiple hits to the head and shake the brain. All of these events can cause concussion or traumatic brain injury that often leads to long-term problems.

Beyond that, this group of brave individuals often witness horrible events involving death, or near-death experiences that scar them emotionally. They try to put what they have seen and felt into a far corner of the brain, hoping not to think of the events.

For the military, and our first responders, their work experiences certainly can cause PTSD. It is my belief that extreme stress such as PTSD can also alter brain chemistry even without a physical injury. This can lead to a large variety of problems both short-term and long-term. It is especially likely that they will experience conditions such as depression, anger, anxiety, and fatigue. Not coincidentally, these are very common symptoms of brain injury and PTSD. A common link may be poor hormone levels.

So, there is bad news. The bad news may come right away with obvious symptoms at the initial diagnosis, or the bad news may slowly become evident over months, years, and even decades.

There is more bad news. Treatment solutions are often poor for all of these problems in many centers around the world.

This book, however, contains good news.

Many times, people are advised to be tough and "shake it off." They have been told that their tests are all fine and so they are fine; there is nothing wrong.

Well, there is something wrong.

More and more, the medical community is recognizing concussions, traumatic brain injury (TBI), and PTSD. Patients are finally being acknowledged with the real problems they are facing.

Many have been told that there is no treatment.

Well, there is.

In this book, you will learn about the new science and treatment for brain injuries and PTSD. You will also see and read about people like you who have not only overcome these problems but have thrived.

Concussions can happen to anyone, no matter your age or stage of life. Are you prepared, if it is you?

UNDERSTANDING CONCUSSIONS

You might be surprised to learn that a concussion does not mean you have been knocked unconscious.

A loss of consciousness indicates a much more severe brain injury. Concussions are only one type of traumatic brain injury (TBI) that can be caused by many activities beyond just sports. Patients at our Komer Brain Institute include people who have had injuries in doing the most common tasks in their normal everyday lives. Sometimes it is a slip and fall at home, falling off a bicycle or often motor vehicle accident.

Of course, injuries in sports are very common. The more intense the contact in the sport is, the higher the likelihood of a concussion. Sports like football, hockey, lacrosse, and even soccer are games where concussions happen. Boxing is perhaps one of the most dangerous sports for concussion-related injuries.

The type of work one does also have variable risks of head injuries from falls, or other causes, including our first responders. Police officers and fire fighters are at great risk. Among those at the highest risk are the men and women who serve in the military. They are often exposed to events that cause concussions and other traumatic brain injuries.

Like snowflakes, every concussion and brain injury seems to have its own unique pattern. There can be a wide variety of symptoms that may or may not be present in each case. The severity of the symptoms will also be different. Some patients have symptoms all the time, while others only experience them during exercises or times of stress. Even changing weather conditions can trigger the concussion-related symptoms.

The typical symptoms of concussion may occur immediately after an event or may be delayed. In some cases, they do not appear for hours, days or even years after the injury.

Research shows that these concussions are accompanied by inflammation of the brain.[1] [2] This inflammation can have serious and sometimes life-changing consequences, as we will discuss later.

Signs and symptoms of a concussion may include:

Physical

- Headache
- Seizures
- Feeling of pressure in the head
- Nausea or vomiting
- Dizziness
- Fatigue or tiredness
- Slurred speech
- Blurry vision
- Light sensitivity
- Difficulties reading
- Problems understanding or remembering what was read
- One pupil of the eye being bigger than the other
- Balance problems

---

1 Rathbone AT et al; A review of the neuro- and systemic inflammatory responses in post concussion symptoms: Introduction of the "post-inflammatory brain syndrome" PIBS.

2 Brain Behav Immun. 2015 May;46:1-16. doi: 10.1016/j.bbi.2015.02.009. Epub 2015 Feb 28.

- Unsteady walking
- Noise sensitivity
- Ringing in the ears (tinnitus)
- Sleeping more than usual
- Sleeping less than usual
- Trouble falling or staying asleep
- Disorders of taste or smell

Emotional or Mood

- Depression
- Sadness
- Anxiety or nervousness
- Irritability
- Becoming emotional
- Personality changes
- Listlessness
- Feeling dazed
- Changes in cognition (ability to think clearly)
- Problems remembering
- Confusion or "brain fog"
- Difficulty thinking clearly
- Brain function feeling slowed down
- Difficulty concentrating
- Challenges in remembering new information
- Delayed response to questions
- Significant loss of memory

A concussion also seems to aggravate pre-existing conditions. For example, if you are a person given to headaches, your major symptom after the concussion may be headaches that are more frequent and more severe. If you are prone to depression, then after concussion, your depression may become more intense.

For those with chronic illnesses such as chronic fatigue syndrome, multiple sclerosis, or anxiety, this kind of event and its stress, tends to make the symptoms worse. For this reason, it is very important to treat the concussion so that the other conditions do not become more severe.

TRAUMATIC BRAIN INJURY (TBI)

The brain can be damaged in different ways. These include infection, tumor, stroke or chemicals, and drugs. However, traumatic brain injury (TBI) is defined as brain damage caused suddenly by an external force.

The adult brain weighs about 3 pounds. It has the consistency of dried peanut butter being both soft and fragile. Cerebrospinal fluid surrounds the brain nourishing and protecting it. The bony skull also protects the brain. However, the interior of the skull has bony ridges that can cause brain damage when the brain is hit by a force that slams it into the skull. Impact to the head that creates rotation can trigger more damage than a direct blow.

Penetrating head injuries occur when an object such as a bullet enters the brain causing damage to a specific area. Closed head injuries caused by a motor vehicle accident, a slip and fall, or in a collision sport, can also cause damage. This type of damage often injures a more diffuse area.

TBIs can result in brain bruising, bleeding within the brain, tears or lacerations of the brain, and damage to nerves caused by shearing forces (known as diffuse axonal injury). Diffuse axonal injury to the area of the pituitary gland may be responsible for hormonal changes seen in a traumatic brain injury.

Concussions described as a "mild" traumatic brain injury is not life threatening. However, the term "mild" can be very misleading in that the initial damage may appear minor even though the long-term effects may be more serious, or even life threatening.

Estimates are that 1.4 million traumatic brain injuries occur each year in the United States, leading to 1 million hospital emergency visits.

Traumatic brain injuries are diagnosed as mild, moderate, or severe, based on the individual's condition at the time of the injury. About 75% of TBIs are categorized as be mild.

With a mild TBI, there may be no visible injury. Even brain imaging (x-rays, CT scans, MRI) shows no brain damage or is inconclusive.

With moderate or severe TBIs, there is typically a loss of consciousness, disorientation or confusion, with neurological signs and changes on a CT scan or MRI. Assessments of these more extreme TBIs are to rule out life-threatening problems such as tearing or bleeding within the brain, or acute swelling. Swelling can cause increased intracranial pressure, breathing difficulties, spasm of blood vessels, lack of oxygenation to the brain, and abnormal heart rate patterns (called cardiac arrhythmias).

Someone who is struck in a car accident, or who falls to the pavement, may injure the brain both at the point of impact and on the opposite side of the brain as well (known as coup-contra-coup injury). This happens inside the brain as it is initially forced forward to the point of impact on the skull and then as it rebounds off the other side of the skull. This injury can cause very diffuse damage because of the tearing effect on the brain, as it moves back and forth within the skull. The Glasgow Coma Scale (GCS) classifies the severity of a TBI. It rates the person's level of consciousness on the scale of 3 to 15 measuring verbal, motor and eye-opening reactions to stimuli. In general, GCS levels of 13 to 15 are classified as mild TBIs, 9 to 12 are moderate and 8 or below are severe. PTSD and a TBI frequently occur together. Among patients hospitalized previously for TBI, somewhere between 35% and 80% have histories of substance abuse, with alcohol being the most commonly abused substance. TBIs are linked to homelessness, panic attacks and anxiety, depression, learning difficulties, domestic violence, suicide and incarceration. Recent figures suggest that 1 million US troops that suffer from TBIs are in jail (perhaps from bad decisions caused by the TBI), and 40 commit suicide each day. A study from Toronto, Ontario showed that 42% of men and 58% of homeless women in that city had significant traumatic brain injuries before becoming homeless. TBIs are not only a health issue for the people who suffer them. They impact families, the workplace, and society as a whole.

PTSD

Post-traumatic stress disorder (PTSD) is common. It is suggested that more than 5 million people United States suffer from PTSD.

PTSD is defined as an anxiety disorder that can occur after a person has been through a traumatic event. Examples of such events can be car crashes, sexual or physical assaults, natural disasters, crimes, or combat during times of war. Typical of these events is a feeling that the person's life, or the lives of others, has been threatened. They feel that they have no control over the situation.

Not all people who go through these events get PTSD. Unfortunately, many do.

Signs and Symptoms of PTSD:

- Anxiety
- Irritability or outbreaks of anger
- Depression
- Nightmares
- Flashbacks
- Sleep disturbance
- Loss of positivity
- Physical responses to triggers of the traumatic event (such as an increased heart rate, a feeling of fear, or sweating)
- Avoidance of any people, places or things that could trigger reminders of the event

It is important to note that many sufferers of PTSD also have had traumatic brain injuries. Recent research by Dr. Dewleen Baker, a psychiatrist at the University of California at San Diego shows that traumatic brain injuries double the risk of PTSD.

As there is such an overlap of symptoms between traumatic brain injuries and PTSD, many of the treatments are similar. Often, patients receive cognitive, behavioral therapy. However, I think that all organic causes need to be ruled out and treated first. This must include an accurate assessment of hormones, with a skilled interpretation of the results. The standard measure of hormones called "normal ranges" by labs is based on levels that include a large segment of the population. They certainly are not ideal or optimal levels; these are just what can be found in the general population. Hormones have to be restored and optimized to best levels not settling for typical levels.

In addition, we know that many supplements can be helpful in treatment. This will be discussed in greater depth in the chapter on supplements. Please note that Vitamin D is particularly useful. It has a major positive effect on mood and energy. The majority of patients I see have nowhere near ideal levels.

In PTSD, it is felt that part of the problem is the inability to form new brain connections. For example, an event or an object linked to the original trauma that caused PTSD may continue to trigger bad memories. The reminder is very different than the actual event, but the body's response is the same. An example of this might be the sound like a siren that happened before a bomb blast. A siren heard long after the original event could still cause a fear response. The ability of the brain to form new memories in response to the siren may be impaired. Magnesium enhances the ability to form fresh non-threatening memories. This allows the brain to be able to discriminate between the old existing memories of the siren and the new ones so that the PTSD bodily response does not occur.

Unfortunately, PTSD is common in our troops, and first responders. I have the greatest respect for the work that they do to keep us safe and healthy. That is why I am particularly delighted when I can help any of these brave individuals.

Much of what is written in the rest of this book is appropriate for treating individuals with PTSD. I want to continue our message of hope. This problem can be treated.

Support groups are also very important. Sharing feelings and concerns with others who have gone through the same situation may reduce anxiety and fears. Family and friends are so important. Understand that those suffering from PTSD may withdraw. Those close to the PTSD survivor need to let them know that they will be there for them. Help them develop a network of family and friends who are willing to listen.

It is important that friends and relatives understand that individuals with PTSD may overreact in otherwise normal situations. They may suddenly become angry, and that anger may seem directed at their family and friends. These times are when caregivers and friends need to step back and realize that the anger is not really about the friend or family member. This is a feature of the illness, and those around the sufferers need to have a high degree of tolerance and patience.

PTSD is as legitimate a diagnosis as appendicitis. We would never tell anyone with acute appendicitis to carry on with every activity of daily living. It makes equal sense that we should never tell someone suffering from PTSD that they should just "suck it up" and get on with their lives.

PTSD is an illness. It deserves our very best understanding and treatment solutions.

POST-CONCUSSION SYNDROME

Almost all the patients I see have post-concussion syndrome. They have been seen by neurologists, psychologists, psychiatrists, physiotherapists and occupational therapists. The reason they have been referred to me is because they are not completely better. As a matter of fact, many of the patients are only slightly improved from their initial injuries. They seem to have "hit the wall" in their recovery.

I have seen 100-page protocols for looking after post-concussion syndrome that make no mention of brain hormones at all. Some of the others suggest that brain hormones should be checked but that is the only reference to them, and there are no guidelines.

Research shows that a very large number of patients with brain injuries have hormonal abnormalities. The pituitary gland and the hypothalamus are responsible for hormone production and function.

Hormone changes should not be a surprise. These organs are located in your head. Your head was hit. These organs are not working optimally now. Correcting the hormones is not rocket science. However, it is good brain science!

The problem here is that many of the traditional professionals looking after concussions, and brain injuries are not experts in Interventional Endocrinology (the science of detecting hormone abnormalities and optimizing them).

As stated previously, it is vital that the medical community understand that some of these early hormone abnormalities are temporary, whereas new hormone problems can become apparent later. Within three months after brain injury, 56% of patients have abnormal pituitary function and abnormal hormone levels. At one year, 36% of patients still persist with poor hormone levels.[3]

When I have lectured to professionals and discussed the findings, most are very surprised. Many are happy to learn that there is another area where the patients may be helped. However, many physicians looking after brain injuries have no training in these areas. It is part of my goal to continue lecturing and educating all professionals about the need for optimal hormone levels for optimum brain chemistry and function, and for healthier happier patients.

The other important new information is that the brain becomes inflamed with an injury. This should come as no surprise. Everyone realizes that if you break a finger, there is going to be a process of swelling and inflammation. However, this inflammation usually slowly heals on its own. The brain is different. The inflammation can continue for years and even decades after the initial injury.

The smoldering coals of inflammation, if unchecked, inevitably lead to the damaging flames that cause cognitive, emotional and neural degeneration.

It may well be that this inflammation makes the brain more vulnerable to greater damage with each successive injury.

What does this mean to someone with concussion or TBI?

3 Journal of Endocrinology Investigation 2005:28 Popovich et al

It means that if you have not had a thorough hormonal workup by an expert, your testing is not complete.

If you have hormone abnormalities (and remember only optimal levels are acceptable), you may still have new treatments, which can improve you considerably.

Having broken hormones is worse than having two broken legs. Nobody would expect you to run a mile on fractured limbs but they cannot see broken hormones. Their expectation is that everything is fine, and you should be able to do everything and feel well. This absolutely is not the case. Broken hormones are as legitimate as broken legs and the damage they do is much worse. Let's fix those hormones.

As we have noted before, if you have not been treated specifically for brain inflammation, your treatment has not been complete. Reducing the brain inflammation improves the toxic environment in which the brain is working. When this happens, the brain has a much greater ability to heal itself. This is known as brain plasticity.

Your brain can still improve!

And so what is the new hope?

We have found two major problems in your brain: abnormal hormone levels and inflammation. These have not been treated. They are fixable.

There is new hope.

GET CHECKED

Right after a brain injury has occurred; the most important medical decision is whether the injury is serious enough to send the person to the hospital to rule out a skull fracture or bleeding inside the head. In these situations, the patient usually gets an MRI, CT scan, or an x-ray.

Most concussions do not require hospitalization. For example, sports concussions are usually assessed on the sideline asking the player some questions to see if he or she remembers where they are, the score of the game, or what happened to cause the injury. This assessment also looks for the immediate symptoms of a concussion such as headache, nausea, dizziness, slurred speech and light sensitivity.

Many athletes have already undergone baseline testing by the ImPACT or SCAT 3 tests. Qualified testers can administer the follow-up testing and see if there are changes that have occurred since the injury. These tests must not be used as the only factor to determine if an athlete should return to playing. That decision ought to be made by a trained professional.

Many patients go home and are asked to follow up with their family physician. What the physician can do is limited, as in most concussions, no abnormalities are seen on MRI or CT or x-ray, even if there are many symptoms.

Although many concussions resolve on their own within days or weeks, many do not. These patients who do not get completely better have post-concussion syndrome. For them, further testing may include neuro-cognitive testing, balance testing, strength and coordination testing, screening of vision, screening for balance or equilibrium (known as vestibular testing), assessment of gait (or how the patient walks), neurological medical assessment or orthopedic assessment for other injuries such as neck, back and shoulders.

Based on the results of tests like these, an assessment and treatment plan can be created.

BEFORE YOU GET BACK IN THE GAME

Return to Play Guidelines

The following guidelines are outlined in the Concussion Census Statement from the 4th International Conference on Concussions.

These steps are done under the guidance of the Sports Medicine doctor and are supervised by a certified Athletic Therapist.

As soon as an athlete presents with symptoms, or we suspect a concussion, the athlete is immediately removed from play and evaluated.

The following steps are done with a minimum of 24 hours between steps. Please note that the athlete must remain asymptomatic for at least 24 hours in order to progress to the next step:

- ➔ Step 1: Cognitive and physical rest
- ➔ Step 2: Biking for 15 – 20 minutes at 70% max effort
- ➔ Step 3: Running/Sprinting for 20-30 minutes
- ➔ Step 4: Non-contact practice. At this point, the student-athlete can also return to progressive strength training.
- ➔ Step 5: Full-contact following medical team clearance
- ➔ Step 6: Return to a normal game

Back to School Guidelines

Here is a helpful link about returning to school following a concussion. It was produced by McMaster University.

https://canchild.ca/system/tenon/assets/attachments/000/000/291/original/MTBI-return_to_school_brochure.pdf

UNLOCKING THE PRISON

Brain injuries can change individuals and their personalities. They can just give you the sensation of not being yourself. You may feel "off." For some people, you may actually feel like you are a different, even an unlikable person. Feelings of instability and deterioration are common after your brain injury. Add to that the emotions of hurt and anger. Life after TBI may make you believe that you are a prisoner in your own body.

People around you may start treating you differently, looking at you in a strange way.

The professionals caring for you may tell you that at some stage you have "hit the wall." They may tell you that at three years (or two years or one year!) after the concussion, you are as good as you are going to get. Sadly, many patients have been told that there is no treatment at all for concussions and brain injuries. This injury has now turned your time in this prison into a life sentence.

As a physician, I am upset and disappointed when any person is treated this way. How dare they take away your hope, your dreams, your confidence, and your self-esteem?

How can they believe that they can predict the future of medical research and treatments? Do they really think that medicine does not evolve and discover new solutions?

A classic example of this is the treatment of stomach ulcers. In the first half of my career, almost every two weeks I could not continue in my operating room because someone arrived at the hospital, critically ill with bleeding stomach ulcers. They needed urgent surgery to survive. I gladly gave up my operating time so that their surgeon could save them. In the last 15 years, this has happened once. Why?

In 1982, two Australian physicians, Dr. Barry Marshall and Dr. Robin Warren, believed that bacteria in the stomach caused ulcers. The then commonly held opinion was that stress and lifestyle factors caused ulcers, and this bacterial theory was ridiculous. They were laughed at by the medical profession. They were treated badly by colleagues. However, they persisted.

In 1984, Dr. Marshall drank a solution containing the bacteria h. pylori. After only three days, he developed ulcer symptoms. A scope of his stomach showed massive inflammation, which could lead to ulcers. They also showed that this condition could be cured with antibiotics.

In 2005, they won the Nobel Prize in Medicine! Yes, it took 21 years for this to be accepted.

So much for the skeptics.

What about brain injuries?

Medicine may be slow to change. That does not mean there are no leading-edge therapies to make sense from a practical point of view, and have good scientific principles behind them.

It often takes years of looking at basic research to give us clues on new therapies. Sometimes it takes looking at your existing medical practice to see how you can apply what you know works to another area of medicine.

That is what I did after years of restoring hormones in both men and women.

A key observation that has provided me with insight into this new treatment of concussions is that a great many of the signs and symptoms of post-concussion syndrome are identical to the signs and symptoms of a low estrogen in women and to the indicators of low testosterone in men. When I restore the hormones, these symptoms get better.

This wide range is not optimal. Many of the patients we have seen were initially told that their values fell into the normal range. Once their hormone levels were optimized, they started seeing improvement. Often, these patients were years and even decades beyond the injury. Improvement was still possible.

I prefer my approach.

Never, ever, ever accept the fact that nothing further can be done.

I want to unlock the life sentence in the prison of despair to which men and women with brain injuries are currently consigned.

HORMONES 101

Hormones are substances naturally produced by glands in the body called the endocrine system. They are manufactured in one part of the body, but used in many other areas. They are the body's chemical messengers.

The hypothalamus is a part of the brain that controls the release of hormones made by the pituitary gland. The pituitary gland is also located in the head at the base of the brain. It is often called the master gland because it produces hormones that signal other glands to manufacture their own specific hormones. These glands include the thyroid, the adrenals, and the ovaries and testicles.

The hypothalamus and pituitary gland are like conductors in an orchestra. The various hormones for which they are responsible are like the many instruments, which have to be in tune, and have to play together. If one instrument is too loud, or too soft, or is not playing with the rest of the orchestra, chaos can result. The same goes for changes in hormones in the human body. They have to be fine-tuned and have optimal levels to work together for the most favorable function and health.

The thyroid gland is located at the front of the neck and produces the thyroid hormones T3, reverse T3 and T4. These hormones control metabolism, breathing, heart rate, mood, muscle strength, menstrual cycles, body temperature and many other functions.

The adrenal glands are located above each kidney. They produce the hormones cortisol, adrenaline, noradrenaline and some of the androgens (male hormones). These hormones have many functions, some of which are the regulation of blood pressure, regulation of metabolism and response to stress.

The ovaries are the female sex glands. Beyond being the reservoir of eggs, they produce the female hormones estrogen and progesterone, as well as a small amount of testosterone. These hormones are responsible for the change in the female body at puberty and the maintenance of the reproductive organs. They play a key role in having periods and also in pregnancy. Importantly, these hormones have a great effect on mood and energy.

The testes or testicles are the male sex glands. Besides being responsible for the production of sperm, they also create the hormone testosterone which influences many areas of the male body, especially mood, energy, sleep, thinking, muscle development, sex drive and bone density.

Growth hormone is produced by the pituitary gland and as its name implies, it is necessary for growth. However, after the growing phase of puberty has been completed, I think it should be more accurately renamed the "repair hormone." It has many effects throughout the body, including maintenance of muscles, reduction of fat, levels of blood sugar, maintenance of good mood, helpful effects on learning and memory, and stimulation of the immune system.

Throughout the day, the body's demands for each hormone change. Your body reacts to what is happening in your life. Hormones are delivered into the bloodstream to cells in various tissues. The hormones bind to a specific cell receptor that is like a key fitting into a particular lock. Once this binding takes place, there is a biological effect. This effect can be either positive or negative. An example of a good effect is the hormone testosterone binding with a receptor in the skin and causing growth of hair on the chin, or its binding to a cell in the testes to produce sperm.

If there is a negative effect, it can stop the production of a substance.

The production of hormones is regulated by a delicate set of feedback loops. This means that the amount of the substance in the system regulates its own concentration. When there are low levels of a hormone, there is positive feedback and more of that hormone is produced. However, when hormone levels get to the appropriate level, there is negative feedback and production of that hormone stops. This is very much like a thermostat on the wall regulating temperature. These balances are very delicate and can be changed by chemicals, stress, and injury.

Since the pituitary gland and the hypothalamus are located near or in the brain, traumatic brain injuries can cause hormone problems. The TBI may change hormone production right away or its can affect hormone levels years, and even decades after the injury. I also have seen many examples in my practice where severe stress greatly affects hormone levels. Post-Traumatic Stress Disorder (PTSD) is a typical example of this.

In summary, optimal hormone levels are critical for health and well-being. Concussions, other brain injuries and PTSD can upset this fine balance and lead to illness and deterioration of health.

However, the hopeful message is that we are now recognizing these hormone abnormalities. We can optimize hormone levels, and those suffering can improve significantly, and some may even get better completely.

RIDING THE ROLLER COASTER

Life is a roller coaster, and certainly your hormones reflect this.

After you were born and while you were young, optimal hormone levels were necessary for you to grow and develop correctly. This is particularly important for the brain. This incredibly complex organ has to develop connections for forming memories and to allow you to learn tasks and skills. The brain also is the control center of the body and maintains vital activities such as programming your heart to beat and your lungs to breathe and your muscles to work.

Substances like growth hormone and thyroid hormone are critical in childhood development. There are many other hormones playing an important role. If these hormones are not optimal, normal development can be a problem.

As puberty is reached, the sex hormones of estrogen and progesterone in the female, and testosterone in the male, are responsible for sexual development. In girls, this causes the development of breasts, pubic hair, the growth of the uterus, and maturation of the ovaries that starts menstrual periods. In the boy, this causes development of the penis, the growth of hair on the body, the deepening of the voice, and the strengthening of muscles.

There are also considerable changes within the brain that lead to the development sex drive and the changes in emotions that occur as adolescents grow. Once this development has taken place, levels of hormones tend to stabilize.

However, in women, this so-called stability still includes monthly changes in hormones. There are volcanic eruptions of estrogen. Blood levels during the day the menstrual period starts are at around 27 pg/mL (100 pmol/L) increasing to 436 pg/mL (1600 pmol/L) at ovulation some two weeks later, and then falling back down over the next two weeks. This 1600% increase in hormones can cause incredible changes in feelings and moods.

In men, testosterone levels are reasonably stable in the healthy individual. They go higher with mild exercise but can be depressed by extreme exercise, such as training for a marathon, or under conditions of stress.

Women come to a time in life called the perimenopause. This is characterized by changing hormone levels. The ovaries are no longer fine-tuned as they are wearing down. They are somewhat unpredictable. Menstrual periods may continue to be regular, or they may start becoming irregular. As early as when a woman reaches her mid-30s, she may complain of menopause-like symptoms like hot flashes and sweats. They are often told this cannot happen, but indeed it does. The symptoms are caused by hormone levels dipping below the optimal range, at some point in the cycle. During the rest of the cycle, the hormone levels may return to normal. In other cases, the hormone levels may shoot too high and cause symptoms of fluid retention and breast tenderness.

As men age, they can see some reduction in their hormone levels. This may exhibit with symptoms of fatigue, sleeplessness, loss of muscle, and mood changes like irritability, depression, or decreased sex drive.

The menopause occurs when the ovaries run out of eggs and hormone levels of estrogen, and progesterone plummets. Menopause is defined as having no menstrual periods for one year. The average age of menopause is 51 with a normal range of 45 to 55.

The symptoms of menopause are any or all of:

- Hot flashes
- Night sweats
- Poor memory
- Poor mood
- Poor concentration
- Poor energy
- Vaginal dryness
- Bladder not working well
- Irritability (the risk of which triples in menopause)
- Depression (the risk of which also triples)
- Joint pain and muscle ache (the only symptom that occurs in 100% of women with very low estrogen)
- Sleep disturbance
- Decreased sex drive
- Palpitations of the heart
- Headache
- Weight gain
- Brain fog (since blood flow to the brain drops 30% and 26 areas in the brain which require estrogen and progesterone are not getting them).

Your menopausal "computer" has 30% less power and 26 fewer programs to run. No one likes this computer!

Symptoms are not the greatest problem. There is an increased risk of serious chronic fatal illness. The risk of stroke and heart attack goes up considerably, with 54% of menopausal women dying of these vascular problems. There are also increased risks of diabetes, bowel cancer, osteoporosis, Alzheimer's disease and dementia and breast cancer.

Restoration of hormones identical to human hormones (bio-identical hormones) reduces these risks considerably. Some calculations put the reduction in deaths each year with hormonal restoration at 29%. This topic will become the subject of a later book we are writing.

Men encounter a similar but not so obvious condition. There is nothing such as the loss of menstrual periods to mark the onset of problems. There are usually much more subtle. This condition is called Andropause or Low Testosterone Syndrome.

Symptoms of low testosterone are:

- Fatigue
- Sleep disturbance
- Loss of muscle and joint pain
- Headache
- Pounding heart
- Hot flashes or sweats
- Development of more breast tissue
- Weight gain around the middle
- Anxiety, anger, depression
- Loss of competitive drive
- Poor memory
- Shying away from social gatherings
- Loss of confidence
- Erectile dysfunction and/or poor sex drive
- Loss of enjoyment of life

As it is with women in menopause, loss of testosterone in men can lead to large increases in stroke, heart attack, diabetes, Alzheimer's Disease, depression and arthritis, among other conditions. These can be significantly decreased with hormone-replacement therapy. This topic is the subject of yet another book that we are writing.

However, what is critical to know is that concussions, brain injuries, and PTSD can depress these hormones and lead to these symptoms in men and women prematurely.

This can happen at any age.

As a matter of fact, many of the symptoms of post-concussion syndrome are identical to the symptoms of low sex hormones in men and women. Restoring hormone levels in these individuals can improve or eliminate the symptoms, improve their health immediately, and decrease chronic illness down the road.

Other hormones may fall as life goes on. In particular, in women, the thyroid hormones are often sub-optimal and this problem needs correction for best health.

BALANCING ACT

Sometimes I feel like a juggler trying to keep a whole bunch of objects spinning through the air. Of course, what I'm doing is juggling hormones, natural products, and supplements to try to come up with unique combination that is the best for each particular patient.

Let's take thyroid, for example.

There are three hormones made in the thyroid gland. The first is T4 (thyroxine). This is a pro-hormone which means it has a little effect on its own, but it is destined to become another hormone. T4 can become T3 (triiodothyronine), which is necessary to produce energy in the cell. However, it also can become reverse T3 (RT3), which has the opposite effect and interferes with the cell's energy production. This is explained by the fact that at times the cell needs more T3 to produce more energy, while at other times it needs to gear down and produces less. That is the function of the reverse T3.

T4 has four iodine atoms, and T3 has three. When a particular iodine atom is knocked off the T4, T3 is produced and the cell makes energy. Reverse T3 is produced when a different one of the four iodine atoms is removed from T4. You can think of this as being like a lock and key arrangement on an engine. T3 is a key which fits into the lock which causes the engine to turn on and make energy; reverse T3 is a different key which fits into the same lock, but does not turn. No energy is produced, and you are prevented from using the good T3 key.

There are four thyroid tests. TSH measures thyroid stimulating hormone made by the pituitary gland to signal to increase the production of T4 and T3. If the thyroid is not responding, TSH levels go higher, whipping the cell to make more thyroid hormone. Low TSH levels mean that adequate thyroid hormone is being produced. In most cases optimal levels of TSH are at the lower end of the range. However, the range for TSH is a very wide. A common range is a 0.35 to 5.0. I feel that the optimal range for TSH is 0.35 to 1.0. I compare this to an eight-cylinder car running on all eight cylinders. When someone gets into the upper end of the range such as 4.99, I believe that this is like having a car running on only two cylinders. The top speed may be a maximum of 10 miles an hour, and that is hardly optimal.

Now let's look at levels of the T3. Modern testing looks at "free T3" and "free T4." T3 and T4 can be bound to a protein called globulin, which acts as a carrier to transfer to other parts within the body. When hormone is bound to globulin, it is not available to be used. When it is no longer bound, it is called free T3 and free T4 and, now it is active and usable. If your body does not convert enough T4 to T3, you will have the effects of low thyroid or hypothyroidism. Similarly, if too much T3 is bound to globulin, you will get the same effect, and have an under-active thyroid. On the other hand, if your body doesn't bind free T3, you will experience the symptoms of hyperthyroidism or overactive thyroid.

Once again, lab ranges are very important. One very common lab range for free T3 is a 3.5 to 6.5. However, with this very important hormone, I feel that your value should be in the upper 25%. This means the ideal range is 5.75 to 6.5. It is not uncommon for free T3 levels to be suboptimal, and you still may have absolutely normal levels of TSH.

The next complicating factor is that when you have an under-active thyroid, most physicians will prescribe Synthroid, which is T4. It is assumed that you will convert some of this to T3. We often see people being treated with T4 who still have low T3 levels and obviously are not converting enough. The answer is to use another thyroid medication that is a combination of both T4 and T3. Giving this ensures that your critical T3 levels will rise. Pure T3 also can be given by itself, but this usually is a much more expensive option.

Now we come back to reverse T3. We can check levels of reverse T3, but I seldom see this done. A common lab range of reverse T3 is 8 to 25 ng/dL in Canada or 90 to 350 pg/mL in the USA. However, this particular hormone should be in the lower 25% of the range which means under 13 (Canada) or under 200 (US). There are treatments for lowering reverse T3, such as giving Vitamin D, giving the trace element selenium (four Brazil nuts per day) or prescribing pure T3.

To make matters even more complex, some people have an inflammatory condition which leads to anti-thyroid antibodies. This can lead to very fluctuating levels of thyroid hormones. There are treatments for this, as well.

Other hormones can affect thyroid function. Increased cortisol can impair the conversion of T4 to T3. Good estrogen levels may improve thyroid function. Many chemicals and drugs can adversely affect thyroid.

So, you can see that looking after just thyroid hormone can be very complicated. Every day I see people who were told that their thyroid tests were normal when, in fact, when I look at them, they were not ideal. Often, only two of the four tests were done. This does not give an accurate picture.

We know that concussions, traumatic brain injury and PTSD can affect all hormones, and thyroid is one that is frequently involved.

When you start looking at all the other hormones besides thyroid, you can imagine the difficulty in sorting all this out.

It really takes an expert in hormones to give you personalized treatment.

You need to have an expert juggler keeping all of those shiny objects in the air at the same time.

This chapter may finally answer the question as to why you are often told that your tests are all normal, but you certainly are not feeling your very best.

Hormone function needs to be optimal.

SHOULD MY CHILD PLAY SPORTS?

This is a very complex question with no simple answer.

There are many reasons to take part in sports.

Children living in this generation are exercising less while living on their computers, tablets, and smartphones more. In the long run, this can lead to obesity as adults. The increased physical activity that comes with sports is important for consuming calories and having a healthy weight. In addition, activity helps burn off energy and gives a feeling of well-being.

Sports can and should be fun! These activities can generate the excitement of competition. Being in sports can teach new skills.

Playing sports also helps integrate children into the real world. Often schools and other organizations try to pretend that children will never fail. We know that this is not the case in real life. Everyone comes up against obstacles that are hard to overcome. People can fail at what they are doing and learn from those defeats as much as from their successes. Children need to be raised with grit and determination as well as love and support. They need to learn to persevere through problems and overcome them. This not only gives them the feeling of achieving success with effort, but it also helps them to learn the valuable skill of problem solving. Sports build character!

Another benefit is that sports teach us how to get along with others, to appreciate worthy competitors, and to respect referees and officials. Sports often lead to developing lifelong friendships. Playing sports when you are young is a significant predictor of your later participation in sports and overall fitness throughout your lifetime. Playing a sport can lead to being an informed spectator giving you an appreciation of the skills of professionals and high-level amateurs.

Sports can teach the child how to deal with difficult people, a skill that is incredibly valuable in later life. These challenges can train the child how to excel under pressure. Engaging in individual and team sports can teach a child how to ignore the skeptics too. Think how valuable this is in the development of self-esteem.

One wise businessman said that part of his selection process in hiring new individuals included asking if they were part of a sports team growing up. He felt that such individuals fit into his organization better, and were generally more successful than those who never played team sports.

However, sports can be dangerous.

We know that the simple act of heading a soccer ball many times can lead to long-term injuries. That is why some soccer leagues are trying to take heading out of the game for children. Football, hockey and lacrosse are collision sports that can lead to head injuries. There are many concussions from cheerleading as these individuals may collide or fall. We know that there is danger in many other sports too.

As children are not fully physically developed and have weaker neck muscles and thinner skulls, they are more susceptible to having concussions. Up until now, many concussions have gone unreported.

However, one has to keep this in perspective. All leagues are now aware of risks and are trying to minimize them. Equipment is improving, and coaching is evolving. This will enhance enjoyment and decrease injuries.

Each parent has to weigh the risk and the benefits of each activity and then make a decision, whether their child should participate. Even the contact sports can be played safely when all involved are educated and aware of new concussion guidelines.

If your child does play a sport make sure that:

- They know how to recognize and to report a suspected head injury
- They know how to explain that such a head injury exists to teammates, parents and coaches
- The child's league or team has a concussion protocol in place
- That you know of a physician who can examine and evaluate your child for the treatment of a concussion. It is vital to have this is in place even before any injury occurs since time is often a factor.

When your child is playing, it is vital that you are very aware of head injuries. You must know the common symptoms of a concussion. These include the following:

- Headache
- Head pressure
- Tiredness
- Drowsiness
- Nausea or vomiting
- Difficulty with concentration
- Sensitivity to light or noise
- Irritability

Remember that although most concussions do not result in a loss of consciousness, they are still a concussion.

Most concussions are mild and typically heal quickly without problems, if the correct approach is taken to their treatment. Supplements may speed recovery. This is outlined in another chapter within this book.

The newest protocols call for resting the brain and the body for days or a week, depending upon the injury, but then returning to physical activity that does not bring back symptoms. There should be a slow increase to previous levels of activity. There should be no full participation in the sport until the injury is completely healed.

There is no question that rushing back to the sport before complete recovery of the head injury endangers the person. It takes considerably less force to cause another concussion. Each successive concussion takes a great deal longer to heal.

There is strong advice to all young athletes to refrain from play immediately when they get injured with a concussion. It is unfortunate to note a new study presented in October 2016 at the American Academy of Pediatrics National Conference regarding the work of Dr. Shane Miller. His research analyzed 185 patients over a ten-month period in 2014 who had been concussed. They aged in range from 7 through 18. The study found that over one-third of the participants ignored medical advice and returned to play on the same day. Previous research indicated that premature return to play could lead to greater injuries, and in some cases even proves fatal, with so-called second-impact syndrome.

This occurs when a concussed person who has not yet healed from his or her head injuries receives a second blow to the head. This can lead to diffuse brain swelling and even death. In the study by Dr. Miller, patients who resumed physical activity too soon reported more severe symptoms than just after they were injured. It is disappointing to learn that even with all the information and publicity about concussions fully 38% of these children were allowed to resume their activity the same day. We still have a long way to go educating everyone on concussions.

If appropriate, school should be notified so that there is a plan put in place to modify activities when the student returns. If the head injury was the result of sports, the aim is returning to health before returning to sports.

On balance, sports can play a crucial role in development. We need to make sure that they are safe as possible and that there are mechanisms in place to ensure future health while learning, growing and having fun.

THE LAST DEFENDER

I like goalies. They are different. They usually get to wear neat looking equipment and padding. They like to have masks which they customize to suit their personalities. They are never happy with their equipment and are constantly modifying it. In soccer, they wear a different (often bright) color to distinguish them from the others on their team. They are often "different." They are part of the team, but at the same time, they are separate. There are times when they perform exceptionally well, and the team still loses. They almost never score on the other team. They often go through times when there is no action, and yet they have to be ready at a moment's notice to make wondrous saves. They are used to criticism, but they learn to let it bounce off them.

For 50 seasons, goaltender was my position. I enjoyed playing hockey from age 3 to age 53, and I have some friends who are still lacing up their skates in their 70s! I admire others who are willing to be between the pipes or the posts. As a goalie, I never felt that I was trying to beat the other goalie in a game; it was their teammates that it was my job to stop. When goalies pass each other on the ice or the floor or the pitch, they often nod or tap each other on their equipment. They are in an exclusive brotherhood or sisterhood. They face the same problems.

In most sports, goalies expect to have contact. It is part of their role in game, even when it is accidental. That means that there is a higher risk of concussion. In this book, you will read that head injuries can lead to other issues, which increase the risk of very serious illness such as heart attack, stroke, diabetes, Parkinson's disease and depression.

Concussions, TBI and PTSD need to be taken seriously. Treating concussions will not only help with the symptoms of head injuries and PTSD, but also may significantly reduce the risk of problems down the road.

This is a whole new approach. I do not want serious problems to happen to you or someone you love.

Together we can make a difference and save lives.

Together we can help families.

Together we can make the world a better place.

Scotty Komer

FIRST RESPONDERS

I have a large number of first responders in my practice. I consider caring for them an honor. This group consists of police, fire, correctional officers and EMS workers. They serve to make our lives safer and healthier. However, they are over-represented in my practice, which is indicative of the fact that large numbers of these men and women suffer from concussions, TBI and PTSD. Their job is often risky and leads to personal injury. In addition, shift-work and stress are huge factors that affect the brain adversely.

On one particular day, I saw two male patients for the first time. They were both 24-year-old police officers who worked on two different forces and did not know each other. However, they were surprisingly similar. Their total testosterone levels were the same at eight nmol/l (U.S. equivalent of 231 ng/dl). These values are very low. As a matter of fact, my father, when he was 90 years old, had better testosterone levels than this. (Perhaps that is one of the reasons that he lived to be a few months short of 100 years old!)

Like many of their colleagues, these police officers were unable to identify whether they had experienced head injuries that may have contributed to these poor hormone levels. Many people have forgotten that they had a head injury. Sometimes they may have had one but only remember the event after talking to their parents who confirm that they did have a head injury as a child.

In retrospect, 90% of the men whom we have seen in the Masters Men's Clinic have suffered a head injury. The aftereffects of the head injury may become apparent years and even decades after it happens.

So, I saw these two healthy-looking young police officers who had horribly low testosterone levels. It really did not matter if the low levels were caused by head injuries, PTSD, stress or shift-work, or a combination of all of these factors. The fact was that they had low testosterone levels that are treatable. This would help their numerous symptoms, especially the fatigue and irritability that they were experiencing. In addition, restoration of healthy levels lowers the risk of stroke, heart attack, diabetes arthritis, and depression later in life. Both were started on hormone replacement and have done very well since.

One of the benefits of my practice is that I have the opportunity to talk with these men and women to understand what their job entails. I feel their eagerness to be helpful to others, even if it means putting their own lives at risk. Every time I see one of these first responders, I thank them for their service, and for making our world safer.

Even though first responders seem unlikely to want to admit they have problems, as a close-knit community, they readily share information with others in their team. One officer explained to me that he was grateful that our treatment made him much better. He was sharper, and that meant he was safer. However, he pointed out to me: "It is great that I am better but if my backup is not, I still could die." Therefore, he was happy to refer a colleague who was suffering.

I have heard the same thing from firefighters and correctional officers. They work together and need to know everyone else is at her or his peak. It is very much like a sports team or a group of siblings. The goal is for everyone to be as good as possible.

When we first started The Masters Men's Clinic, Gord Tonnelly, our Clinic Director, and I provided 10 educational sessions over a period of two years at a local steel mill. This company was very advanced. They frequently had health and safety days for their workers. The education spread by word of mouth throughout the mill. Many of the men became patients. They told us the men all talked openly about their problems and about how much better they became with treatment.

This showed us that education could change a culture. Steelworkers are tough, and they don't complain. However, they also were concerned about others and spread the good news.

One of the reasons for writing this book is to educate more people. I would hope that you would pass along new information and the message of hope in this book to others in your lives. It seems that almost everyone we talk to have a friend or have a relative who is suffering from concussions, TBI, or PTSD. Many are struggling with their recovery.

Perhaps you will be the first step in helping them improve.

Badges

OUR UNSPEAKABLE LOSS

Kimberly Dawn Komer

We were so thrilled when Kimberly Dawn was born. It had been a stressful year of infertility.

Little Kim sang nursery rhymes constantly and always perfectly on key. It did not matter where or when. Whether she was in her bed or she while she was eating, there was always a song. Later, she sang in musicals and choirs at school and church. Kim loved to perform as a dancer, singer or actor. She was a triple threat.

Academics was a serious business for her. Whatever she did she excelled. Graduating from High School as an Ontario Scholar, she earned many scholarships. Working hard, she paid her way through the four years of her Honours Arts program at Queens University. Kim was fiercely proud of her independence.

Books were Kim's special friends. Well ahead of typical reading for her age, she had her own treasured Grant's Anatomy book.

Our wonderful, Kim was a beautiful loving spirit. She was sensitive. As a very young child, she seemed to be tuned into the true vibrations that people gave off which eluded many of the rest of us.

Animals were one of her loves. As an adult living on her own, she rescued a rabbit and a dog and gave generously to shelters and rescue organizations. Boxers became her dog of choice; Sierra and Lincoln were her cherished babies for six years.

Kim had a way of commanding a room the minute she walked into it. It was a trait that her workmates loved about her. She also had a wicked sense of humor. In all the pictures of her early life, she was always smiling that big ear to ear grin. She had no EGO, unusual for an attractive woman. Part of her charm was her self-deprecating humor. She didn't mind making fun of herself, telling a funny story about something goofy or embarrassing that she had done.

When she finished her undergraduate degree in social science at Queens, in typical focused Kim fashion, she decided Human Resources was a sound area and enrolled in the Master's Program in Industrial Relations, at the University of Toronto.

Her first job in Human Resources, during the lean and mean late 1990s, found her mostly firing people. The experience prompted her to make a change to pharmaceutical sales.

With no science background, she was not likely to be hired. However, an enlightened manager at Schering Pharmaceuticals took a chance on her. While she had no sales experience, she was a natural performer, with a great outgoing personality. She became very successful earning awards as a pharmaceutical representative.

In her first week on the job, she attended a large offshore meeting, still not knowing most of the other members of the company. In those early days, during an evening of entertainment by singer Corey Hart, she volunteered to accompany him and lay on the piano as they sang a duet. She later told us it was the only Corey Hart song that she did not know! She still pulled that off and made quite a first impression.

Kim was tall and beautiful. She was smart, sensitive and loving. Even with so many gifts and accomplishments, she was constantly suffering from low self-esteem. We could never understand why that was so. She excelled at everything she did. She was an outstanding salesperson, as evidenced by her growing bank account, yet she felt incompetent in the role. We could not understand her constant anxiety.

In May 2015, Larry attended the first ever conference on hormones and traumatic brain injuries given by Dr. Mark Gordon. Mark had been a dear friend over many years before this and was instrumental in raising our interest in the way traumatic brain injuries affected hormones. Mark's 2007 book on Interventional Endocrinology had a chapter on Post-Traumatic Brain Injury Syndrome, and soon after that we started restoring hormones to the brain injured in the practice.

On that day in May, Mark was talking about the large number of brain-damaged U.S. troops coming back from Iraq and Afghanistan. In the presentation, he mentioned that many of them developed a scar on the brain. A light suddenly went off for us.

Kim had developed a scar on her brain when she was 18. Many doctors looked after her, but none could discover why she had that scar. She developed seizures and later depression. Mark pointed out that the scars were almost always due to a traumatic brain injury.

Driving home from a late-summer concert, her car was T-boned and destroyed by a drunk driver. We were so grateful that she had not been "seriously hurt," aside from some bruises. Unfortunately what we did not know then, the T-bone injury caused a bleed in her brain that began a series of brain changes. The result was aggravated anxiety and depression that never responded to any anti-depressant medications. Having just started University, she was away from home, so we weren't aware of what was going on. We only knew that she was not very happy.

In addition, what Mark was finding in the injured troops was ongoing brain inflammation. If not treated, the condition led to a continued downward course, as we had seen in Kim.

We did not know that alcohol became her little secret way to cope. Substance abuse is common in this type of injury. On the surface, she dealt pretty well most of the time, and she was very good at keeping it secret.

Later, it was a miscarriage then a divorce. A move and job change brought her to her knees. She never seemed to be able to get back on her feet. She was in and out of hospitals. She was in and out of rehab. Her abuse of alcohol increased. She was continually unhappy.

Kim's strong will and her independence, kept us at arm's length. Typical of those battling addictions, she lied to herself and to us. Our beautiful superstar was living in squalor, unable to do even the most basic things for herself and her beloved dogs.

Kim passed away on May 7, 2015. She was only 40 years old. Larry was in California, just one week later where Mark was sharing ten years of research into the science of brain injuries. Larry said, "Here I was in California, grieving my daughter, missing my wife and son who were going through the same terrible emotions at home, when what Mark was saying hit my like a sledgehammer."

What we had gone through with Kim finally made sense. All the anger and guilt and frustration with her that she could not get her life back on track, disappeared with our new understanding of brain trauma. She was the innocent victim of a drunk driver's recklessness that caused her teenaged brain damage; because it was not known, and therefore not treated, Kim went on to follow a long, slow and downward slide until her death many years later.

It was too late to help our Kim, but we are determined to help save others by writing this book so that their families and friends do not suffer the same heartbreak that we have.

We have three important messages for anyone reading this book.

First, early intervention after a TBI or concussion, as we have described, is much more likely to stop the resulting terrible cascade of damage, physical and psychological, from the inflammation in the brain.

Second, by sharing our dark family secret, we hope that everyone will realize that this can happen to anyone. We kept telling Kim that if she kept on her self-destructive addictive ways, she was going to die. She did not believe us, as most do not. She did die.

Third, and most important is for years we suffered a lot of embarrassment and regret and terrible guilt. Now, with a new understanding of brain trauma, we realize that it was not our fault, that it was not something we did or neglected to do. This has been the most wonderful gift.

We hope we have been able to pass it forward to you.

WALTER
THE
LAZY MOUSE

KIM'S SONG

Music & Lyrics by
Joan Chandler Komer and Scott Komer

Chorus

You couldn't live in this world and we never knew why.
Your pain was so great it was easier to lie
To yourself and to those who had loved you so much.
Couldn't live in this world. Your pain we couldn't touch.
And where does it come from?

Verse 1

A baby you're born and all that you know
To love and be loved discover and grow.
You're going to be great you'll make your own path,
The brightest of lights you'll never look back.
And somewhere along it slithered into your life
It deceived . The darkness it grew.
It lied with such ease.
It knew what to do.
Your sworn enemy
Was killing you.
What's killing you?

Chorus

You couldn't live in this world and we never knew why.
Your pain was so great it was easier to lie
To yourself and to those who had loved you so much.
Couldn't live in this world. Your pain we couldn't touch.

Verse 2

A Merry go round of emotions and fears..
Days into months, years into tears.
Sometimes enabled and sometimes withdrew.
What did we do? What did we not do?
Understand now, believe the disease
That grabs by the throat and cuts to the knees.
We watched and we hoped and prayed for a guide..
You couldn't be honest with yourself or the world.
There was only one outcome and that was assured.

Chorus

You couldn't live in this world and we never knew why.
Your pain was so great it was easier to lie
To yourself and to those who had loved you so much.
Couldn't live in this world. Your pain we couldn't touch.

Verse 3

So now you are free from the pain and the sorrow,
For us there will never be another tomorrow
Including your smiles and your hugs and bright future.
With sadness and loss we shake fist at cruel nature.

Angels now have you in tight embrace,
The closing night crowd to see your shining face.
And now that it is over we have to believe
We'll see you someday, and we know you're at peace.

Final Chorus

You couldn't live in this world and we never knew why.
Your pain was so great. It was easier to lie
To yourself and to those who had loved you so much.
Couldn't live in this world. Your pain we couldn't touch.
Couldn't live in this world and we never knew why.
Your pain was so great. It was easier to lie
To yourself and to those who had loved you so much.
Couldn't live in this world. Your pain we couldn't touch.
Your pain we couldn't touch.

We wanted to express, somehow, our pain and frustration, of having a loved one who could not conquer her addictions, even when they were killing her. So Joan and Scott Komer, mother and son, co-wrote a song called, "Kim's Song." Accompanying himself on guitar, Scott sang "Kim's Song" at Kim's memorial.

The words are included within this chapter, with the link to the performance sung by Scott are on our websites, including www.drkomer.com.

We hope reading the lyrics, and hearing the song, will help those of you who have had a similar experience.

We also hope that if you are the loved one battling your demons, these words will help you understand the impact of your choices on friends and family. No one is an island.

However, we want to point out that the song was written after her death, but before we came to understand why she became so sick after her head injury in the car accident, which had happened so many years before.

Now, our grief is not less, but our anger and guilt are gone, replaced by pure love.

A STORY OF HOPE

Each day at my practice, I look forward to the positive updates from patients that we have helped through their challenges. They have stories that remind us why there is hope and why these opportunities to help really matter.

When I first meet a patient, we usually set a goal together for our work together

Often, they are:

- I want to get back to work
- I want to be able to go back to my sport
- I want to walk my dog
- I want to help others with my problem

At night, Joan always asks me to tell her "The Story of the Day."

Here is one of those stories.

James is a 31-year-old man who I saw at the Komer Brain Injury Institute recently. This was at the urging of one of Joan's friends, Sue Paquette, who is a superb caseworker for such patients.

He was very unhappy and stated that there probably was not much I could do since no other doctor had really helped him. His mood was more down than up and this certainly was a down day.

Involved in a serious motor vehicle accident in 2008, he had a long history of seeking help. James saw 15 physicians before he came to the clinic. As it often happens, he had been assessed at the local Acquired Brain Injury Clinic but was not helped.

When we assessed James, he scored 203 out of 250 on the Masters Andropause Screening Index, a 50 question survey outlining the symptoms of low testosterone developed in our Masters Men's Clinic.

This is in the worst 1% of all scores we have seen in the more than 6000 men we have treated. He was suffering.

He also scored 50 on the Beck Depression Index II which corresponded to extreme depression. On the Rivermead Post-Concussion Symptom Questionnaire, he scored 8 on the symptoms. This is usually seen immediately after concussion and 43 for symptoms seen later. This once again, is an extremely high score.

My usual protocol to decrease inflammation in the brain and allow it to heal finally, includes optimal doses of Vitamin D and Omega–3 fish oils. Later lab investigations showed his testosterone level was 1/3 of the ideal. I started him on testosterone replacement therapy.

I saw him for the six-week follow-up. When asked how he was feeling, he grumbled "maybe I'm doing a bit better." Caseworker Sue looked at him and invited him to share his story.

"My dog got sick and I wanted to help him, but I did not have $4000 for surgery. A friend of mine offered to put up the funds and told me I could pay him when I got some money. I looked around trying to find a vet who would do the surgery cheaper. I found one for considerably less and told my friend. He said to go to the original veterinarian and have the surgery done there. He said he felt the first clinic was better and told me I would only have to pay him back the amount that the cheaper clinic would charge. He would pick up the difference."

"My dog had the surgery and did very well. I was so grateful to my friend that I wanted to do something for him."

All of this is coming from a man who has not worked for eight years. He had not been able to do anything resembling his previous occupation as a framer. James was very depressed in the past and barely functioning. He also had been given the usual story by professionals looking after him that he would not get better.

Then he said that his friend needed some work done around his patio.

He told me, "So I did this for my friend."

James showed me a picture of his work. I blinked and choked up as I tried to process what I was seeing. The picture was of an incredible structure he built on his own in just a few days - without any blueprints!

I told him that I could never build this in 100 years. Standing up, I reached across my desk and asked him to shake my hand. I told him "You are incredible! You should be very proud of yourself"

He admitted: "I am not feeling so worthless anymore. I am proud of myself."

Here is a man who was told he would not get better. Like many others, he came to accept that this would be his life. Decreasing the brain inflammation and restoring his hormones to the correct levels brought out the real man buried underneath the problem. He was incredibly tough; to fight given all he has gone through for the last 8 years. He will thrive and be much closer to his old self. James has his life back.

Good friends are also essential on a painful journey. I commend his friend who offered to pay for his dog's surgery. There are good people in the world, and he is one of them.

James wanted me to share his story so that if you are one of those people who are weary and hopeless on the journey, you too must have hope.

Another day at the practice is another day of one or more stories to tell Joan of despair turned to hope.

What a privilege it is to be part of their recovery.

THERE IS NOTHING MORE WE CAN DO

Cody is an awesome young man whose story needs to be told. I think of him every day in my clinic. His story recharges my energy to help other patients suffering from concussions, TBI and PTSD. Reading this may change your life as well.

His story is typical of several of the patients we see with concussions and PTSD. Such patients have often spent months, or even years seeing health practitioners in their quest to get better.

A male patient I met recounted the 40 physicians and therapists he had already seen, before talking to me. These patients are referred to me because they are still not feeling well and usually are in great distress.

The typical journey most patients have followed often includes seeing a combination of specialists. This may include a neurologist, a psychiatrist, physiotherapists and occupational therapists, psychologists and behavioral specialists, a family physician and in the case of a brain injury, an Acquired Brain Injury Clinic.

Thankfully, many patients get better with this help but far too many do not. Our clinic has a different protocol for treating these injuries. We usually see people, who in spite of all these typical treatments, continue to suffer. These patients are caught in a whirlpool that is spiraling downward with no next step. Depression can set in, and life for the patient can get worse and worse.

Cody* is a 24-year-old former hockey player who had received a concussion three years before I met him. During those years he "hit a wall" in his recovery. He was referred to me by his family physician who very wisely checked his testosterone levels and found that they were low. He had started on testosterone and had a slight improvement before I saw him. His mother accompanied him to our first consultation. Cody told me his story.

He played hockey at an elite level. Now, he not only could no longer play hockey, but was not functioning well. Often, he was tired and did not sleep soundly. There was loss of muscle and decreased range of motion. Cody suffered from hot flashes and definite mood changes, including being easy to anger, feeling burned out, having a loss of competitive drive and shying away from social gatherings. While he had improved a fair bit from the testosterone injections, there was still a long way to go for him to feel well. He had concerns about the therapy.

I told him that I could understand his position, since I deal with many athletes in my practice. Just the inability to be able to play your beloved sport is a tremendous loss. Add to this the symptoms of post-concussion syndrome, and there is a substantial decrease in enjoyment of life. I let him know that I saw others in exactly the same position. I was confident that with adjustments to his therapy, Cody would be functioning and feeling considerably better within six weeks.

Good news like this was not something that he was used to hearing. He stood up and gave me a hug. Seeing someone like Cody discover that there was new hope is incredibly satisfying.

Then, Cody sat down and told me that four of his friends had undergone similar injuries and struggles. I was about to tell him that I would be happy to see them as well. Then he said, "All four of them have committed suicide."

I have heard many difficult stories in my 40 years of practice, but this single statement hit me like a tidal wave. I was totally unprepared for it. I asked Cody if he had ever considered ending things, and he affirmed that it had crossed his mind too. Then he turned to his mother and said, "But with parents like mine, I would never go ahead and hurt them."

I put together a treatment plan and said I was really looking forward to seeing him in six weeks' time to update on how he was doing. When Cody kept that follow-up visit, he came in with a smile. He was feeling much better and was happier. His low mood was improving.

Cody told me that he had written the story about his troubles on Facebook. He outlined how he had finally recovered so much of what he had lost. He had over 800 replies. Cody contacted many of them to reassure them.

I had the pleasure of seeing Cody and his mother recently. He is now off antidepressants. He is taking over the marketing role in the family business and sales have improved significantly. Now he says that his brain is working better than before his injuries. I think that indicates that he had some previous concussions before the major ones in Junior A hockey and, already was having some minor problems before that.

Cody has offered to share his story by even speaking to groups of his troubles and about how he feels very much better. It has been an honor to look after such a fine young man who now has so much to offer the world.

Before, he was in a downward spiral, caught in the whirlpool of despair.

Now he is free and is soaring to new heights!

*Cody asked that we use his real name in telling his story.

DOCTOR, I HAVE THIS FRIEND...

How should you talk about your injury with your healthcare professionals?

This is a difficult subject. Let me give you some tips from the physician's point of view, so that you can maximize the time and success of your visit.

It is very helpful to bring a concise written summary of the following information.

We want to know your history of head injuries or the traumatic events that led to the diagnosis of PTSD. This should include dates of the incidents.

Next, include a brief medical history of other serious conditions in your past, not related to the head problems. Do not waste time trying to list everything that has happened to you in the past. A shopping list of your life history slows down and confuses the interaction between you and the doctor. It takes away valuable time that would be better spent in trying to put together a comprehensive treatment plan. If the physician wants you to elaborate on the list of problem or wonders about other specific things that you have not included, he or she will ask you directly.

Then make a list of the symptoms and problems that you had initially, and indicate the ones that are still ongoing.

Next, have a list of the physicians and health professionals that you have seen for your head injury or PTSD.

List the medications or treatments that you have had, with the approximate times that these occurred. Identify whether they were totally or partially successful, or were of no benefit.

Remember that you are dealing with a very busy professional, and you want to make this information concise and accurate. This will allow maximum time for the important steps of your physician making a diagnosis, outlining tests, talking about treatment options and discussing success rates.

What you do not want to do is continue discussing the past, restating how poorly you feel now. This often hijacks a conversation at a time when you need to be thinking about the future, looking for ways that you can improve your life and health.

Do not be afraid to ask questions during your discussion about tests, treatments, and results. When all of this is done you may want to discuss some investigations and treatments that you have not had which, perhaps, the physician has not offered. I would hope, as you learned from this book, that you would ask about hormonal assessment. In addition, consider the assessment of visual function, assessment of dizziness, hearing and balance, if this is an issue. You may also ask about management of pain or stiffness, particularly in your back and neck. Correcting all of these problems plays a critical role in returning to optimal health.

The other important thing is stating your goal. Do you ultimately want to get back to work? Do you want to resume playing your sport? Perhaps, you want to be able to face social situations more easily? Has your illness caused real difficulty in relationships and is solving this major goal? I often write down the goal that the patient has in my chart and highlight it, because this not only gives both of us something to work towards, but also allows us to recognize when success has been achieved.

Often the goal of the patient is very different than the goal of the physician. These differences need to be brought into line.

The importance of goal setting should not be minimized. I feel that patients should not only write a goal in a positive manner but also put a time frame on it.

Just having a goal of "getting better" is too vague. Research has shown that goals are much more likely to be achieved when they are stated in a positive manner and also by including a date by when you hope to achieve them. Chances of success are higher when a mood or emotion is expressed. Be realistic about the goals you set, so that you have a higher chance of success.

Let's say you want to get back to your sport. You may state your goal as, "I can see myself happily playing hockey with my friends by September of this year."

What you have done is visualized the goal, included an emotion and stated a time frame.

A difficult problem develops when you ask a physician about a treatment with which he or she is not familiar. If you do this in a non-threatening manner, many physicians will at least consider the possibility of including these in your treatment. If the physician is not familiar at all with the treatment or investigation, the good ones will admit that they do not know, but they would be happy to refer you to another expert for that assessment.

A disappointing situation occurs when the physician fails to acknowledge that your line of thinking has any merit at all, and dismisses it completely. I hope that this would seldom be the case. However, many of the patients I have seen have been in this position.

Two patients impressed me substantially with their insight.

One stated: "It is very difficult to change the mind of an educated person." You would hope that the opposite situation was true but, sometimes it is not.

Another much younger patient of mine used the language of her generation when she told me, "Doctors are down on what they are not up on." I told this teenager that she was brilliant to phrase it this way. I have never forgotten her words of wisdom.

As a physician, I am constantly hearing from patients about treatments that they have read about it, or learned about from their friends. I hope that I am always tolerant about considering what they say. If I have information to show that these treatments are not scientifically valid, I will explain to them the evidence that I have. In my younger days, I quickly dismissed ideas that did not jive with what I had learned. However, as time went on, I saw that the gold standard (the ideal treatment) frequently changed to something else. In fact, like a soup on a menu which changes daily (the soup du jour). The gold standard may just be the gold standard du jour. I try not to be locked forever into the information I have today.

New methods or ideas are regularly dismissed quickly because they seem ridiculous because they challenge what we think we know. However, eventually, with time they may be proven superior.

The key message I want to leave you with is this. There is always something else out there that may improve the way you are. Doctors are continuously searching for new methods and innovative treatments. Never give up hope. Always believe that there is something more that can be done. Do not get so beaten down that you feel no further improvement can take place.

So many of my patients are in this position. When I see them they have not had their hormones checked. I have so many examples of significant improvements from people who have seen a multitude of health professionals before me. I don't consider myself unique. There are countless excellent physicians who want to help you.

Keep looking and stay positive. When you do get better, share your news with others. There are so many going through what you are going through.

Share hope with them.

THE ONE AND ONLY YOU

Increasingly, medicine matches treatments to the unique characteristics of your health profile. This customization of medicine reflects the success of research, showing the variations of how the same health experience may vary in the way it presents in different people. We are all unique. This fact matters when it comes to choices about your health and wellness.

As you would expect, medicine still begins with the generalization of what symptoms or signs fit a known condition. Most treatments and protocols are still applied generally to anyone fitting the general profile. That allows for the limited resources of the healthcare system to be applied most efficiently.

What is new are additional questions now being included the process. Is there anything about this particular disease, condition, or injury that may have unique characteristics for this individual that might be different from the general treatment plan? Put another way, what can we do to get the maximum benefit for this patient from the treatment options that are available?

Some areas of medicine do not yet have many ways to customize treatment to the personal health signature of the individual. As we have seen in other areas of medical research, these "one size fits all" approaches will begin to disappear as we better understand the genetic differences, and individual complexities that makes each of us unique.

In the area of brain health, we now understand that patients need to be diagnosed and treated as individuals. No two concussions are the same.

The same physical injury may impact one individual very differently from someone else, depending on other aspects of their overall health and how they respond to the trauma. Not everyone with a concussion will display all of the same symptoms all the time. Some symptoms may appear immediately in one person, and not show up in another for weeks, months or even years. As mentioned in an earlier chapter, you may have had a concussion, even though you did not lose consciousness. Each experience is unique. That should be assumed, especially when it comes to such a complex organ as the brain.

Treatment plans for concussions, traumatic brain injuries and battlefield traumas need to match both the nature of the injury and how it impacts each individual, in particular. In the short term, it may seem more efficient to create a simple "this for that" method that would apply to everyone. However, the most efficient approach to medicine in the long run is to have each patient experience optimal health. Individuals who are not able to function to their potential tend to have more health challenges later. They also do not enjoy the quality of life possible, thereby robbing them of life experiences and the ability to contribute fully to the community. So, a customized treatment that matches the particulars of that individual is better than a general answer that ignores our differences.

This is especially true in any hormone therapy that is applied as part of the treatment protocol. Identifying the plan for optimal results is a key part of the process. Ongoing research combined with clinical experience allows health professionals to maximize the good outcomes for each individual in their care.

When it comes to finding help for the treatment of a traumatic brain injury, battlefield trauma, or concussion, ensure that your health professionals explore the unique aspects of injury your love one has experienced. It may take a team of healthcare professionals to help you resolve the many related challenges that often accompanying a head injury.

Understanding those differences and matching a treatment plan that is customized to you can improve your prospects for recovery. That is better for you. It is also better for all of us who are in community with you.

BRAIN FOOD

In addition to restoring ideal hormone levels, we need a number of supplements to optimize our maximum brain recovery potential and ongoing maintenance. The conversion of each hormone requires specific enzymes which in turn require specific vitamins and minerals. As a result, deficiencies of vitamins and minerals can play a significant role in hormonal balance. Some medications affect vitamins and minerals. Your pharmacist might be a good resource in this regard.

The reliance on pharmaceutical products instead of natural products for similar improvements is almost out of control. One of the benefits of natural supplements is that most have fewer side effects than drugs. However, remember that the effects (and side effects) of each supplement must be evaluated carefully both before starting and when continuing treatment.

A major problem with natural products and supplements is the great variation in quality control. While commercial drugs undergo strict guidelines for safety and content, natural products are nowhere near as carefully evaluated. Careful review of the great number of brands of each type of supplement is part of choosing your course of treatment. I am working with a couple of organizations, including a well-known university, to identify products with the highest quality control and efficacy. Updates on this will be available on our website and in our social media feeds.

Vitamin D

Excellent Vitamin D levels are necessary for brain healing and brain maintenance. Optimal Vitamin D levels slow brain aging. I optimize Vitamin D levels for acute concussions, post-concussion syndrome, acute TBI and PTSD. Vitamin D easily crosses the blood-brain barrier. This barrier is a membrane that keeps any unwanted substances out of the brain and allows helpful ones to pass in. Vitamin D is critical for brain function. Excess oxidation and inflammation disturb good brain function. Vitamin D acts beneficially as both an anti-inflammatory and an antioxidant in the brain.

Vitamin D is not really a vitamin but is a hormone. The definition of a hormone is a substance made at one site but used at another, and this is exactly what happens with Vitamin D. It is made in the skin and used throughout the body.

It is yet another critical hormone to optimize.

Two-thirds of the US population has sub-optimal Vitamin D levels; the percentage is probably even higher in Canada because of fewer days of sunlight throughout the year. Less than 5% of the patients I first see in my office have ideal levels.

The blood test to check Vitamin D levels is "25 – hydroxy Vitamin D." Once again; laboratory ranges are completely misleading. Each lab often indicates different optimal ranges. One such range is 40-260 nmol/L. This is incredible since the upper end of the range is 6 ½ times greater than the lower end. Ideal levels are around 150 nmol/L (60 ng/mL).

Low levels of Vitamin D are associated with disturbed brain function, a number of cancers, heart disease, and obesity.

Vitamin D is crucial for brain function. Depression has been associated with low Vitamin D. Seasonal affective disorder (SAD), which cause people to become depressed during the winter, may in large part be due to lack of sunshine and poor Vitamin D levels.

Vitamin D is also important for energy production.

As we use more sun block, people make less Vitamin D. As we age, we make less Vitamin D. By age 65, Vitamin D production is seriously compromised.

Most of the patients in my practice are either not on any Vitamin D, or they are taking 1,000 or 2,000 international units. This seldom is enough. I often start them on 5,000iu per day and then titrate the dose to get optimal levels. Vitamin D is available as drops, tablets or gel caps. I prefer the gel caps because of better absorption Take tablets or gel caps with food with some fat content for good absorption. Even butter on toast will do.

At the higher blood levels, there is a concern that Vitamin D may lead to calcification of blood vessels and heart valves. For this reason, I usually suggest a supplement with vitamin K (a combination of both vitamin K1 and vitamin K2) of 2500 mcg daily to offset this effect. Patients with clotting disorders should check with their personal physicians regarding vitamin K.

Omega-3 Fish Oils

Omega-3 fish oils enable brain healing and a slow brain aging.

I optimize Omega-3 levels for acute concussions, post-concussion syndrome, acute TBI and PTSD.

The brain is composed of 60% fatty acid. The Omega-3 fish oils contain large amounts of the fatty acids EPA (eicosapentaenoic acid) and DHA (docosahexaenoic acid).

EPA and DHA cross the blood-brain barrier. These fatty acids are anti-inflammatories and antioxidants in the brain and help decrease the inevitable inflammation that occurs after brain injury. They improve cognitive function, memory, and focus.

Besides their value in brain healing and brain function, EPA and DHA also improve heart health, reduce triglyceride levels, and are helpful in inflammatory disorders such as arthritis. These oils increase protein synthesis that helps maintain muscle. They have beneficial effects on the immune system, and this may reduce cancer rates.

You can increase your Omega-3 levels through foods such as wild salmon, hempseed, walnuts, and avocados. However, many people do not get enough Omega-3 in their diet and must use supplementation.

I usually recommend 3,000 mg of Omega-3's taken daily. To get this amount I ask my patients to look on the label of the supplement they are using. This will show the EPA and DHA content per capsule or teaspoon. They then take enough to reach a combined total of EPA and DHA of 3,000 mg per day.

If you are going to take this or any supplement yourself, I always recommend that each person consult a knowledgeable healthcare professional about their unique situation.

There is a tremendous variation in the quality of Omega-3's, perhaps more so than any other supplement. There are various processes used to refine the Omega-3's and they are not all equal. I am now consulting with an extremely knowledgeable University department to find the very best product. I cannot stress strongly enough that the omega-3's need to be of the highest quality (Pharmaceutical Grade Omega-3 Oil). See our website for details. In addition, we will soon have available a test to determine optimal blood levels of Omega-3. See our website for details.

CURCUMIN (Turmeric)

I use curcumin to treat acute concussions, post-concussion syndrome, acute TBI, and PTSD.

Curcumin is a substance in turmeric, which is known as a spice and is one of the main components of curry powder. There are smaller amounts in ginger. It is a potent antioxidant and anti-inflammatory. These properties may be helpful in reducing the damage after brain injury. Curcumin may reduce normal brain aging and reduce the cognitive decline.

The dose of curcumin depends on the specific product and other substances in the product that improve absorption. I usually recommend 400 mg of a bioactive curcumin daily, as it is the most potent form. Larger doses of conventional extracts are necessary to achieve good blood levels.

Taurine

I use taurine to treat acute concussions, post-concussion syndrome, acute TBI, and PTSD.

Taurine is a strong brain anti-inflammatory.[4] Taurine is an amino acid that plays an important role in creating new brain cells.[5] [6] [7] Growth of new brain cells in the hippocampus area of the brain improves cognition

4 Ward R, Dexter D, Crichton R. Ageing, neuroinflammation and neurodegeneration. Front Biosci (Schol Ed). 2015;7:189-204

5 Gebara E, Udry F, Sultan S, Toni N. Taurine increases hippocampal neurogenesis in aging mice. Stem Cell Res. 2015 May;14(3):369-79

6 3Pasantes-Morales H, Ramos-Mandujano G, Hernandez-Benitez R. Taurine enhances proliferation and promotes neuronal specification of murine and human neural stem/progenitor cells. Adv Exp Med Biol. 2015;803:457-72

7 Kim HY, Kim HV, Yoon JH, et al. Taurine in drinking water recovers learning and memory in the adult APP/PS1 mouse model of Alzheimer's disease. Sci Rep. 2014;4:7467

and memory.[8] [9]

Taurine levels fall normally with age. Supplementation of taurine may help the heart and other organs, as well as the brain. Using Taurine demonstrates a special benefit in both Parkinson's disease[10] and depression.[11]

My patients use taurine 1,000 mg twice a day for improved brain health.

Magnesium

Magnesium is critical in enabling brain plasticity, a condition where the brain can change and heal itself.

I use magnesium to treat concussions, post-concussion syndrome, acute TBI, and PTSD.

Magnesium is a key substance to optimize brain function. Brain injuries often result in difficulties with memory and learning. Magnesium is a key element in improving both of these functions. It works with calcium to cause the release of neurotransmitters necessary for normal brain activity. Magnesium deficiency slows the brain's recovery from injury. Studies suggest that optimal levels of magnesium can reverse the aging of brain cells.

---

8 Kim HY, Kim HV, Yoon JH, et al. Taurine in drinking water recovers learning and memory in the adult APP/PS1 mouse model of Alzheimer's disease. Sci Rep. 2014;4:7467

9 Wenting L, Ping L, Haitao J, Meng Q, Xiaofei R. Therapeutic effect of taurine against aluminum-induced impairment on learning, memory and brain neurotransmitters in rats. Neuro Sci. 2014 Oct;35(10):1579-84

10 Perry TL, Bratty PJ, Hansen S, Kennedy J, Urquhart N, Dolman CL. Hereditary mental depression and Parkinsonism with taurine deficiency. Arch Neurol. 1975 Feb;32(2):108-13

11 Zhang L, Yuan Y, Tong Q, et al. Reduced plasma taurine level in Parkinson's disease: association with motor severity and levodopa treatment. Int J Neurosci. 2015 May 23:1-24

Not all forms of magnesium are equal. Most do not increase brain levels significantly. The best form of magnesium that increases brain levels is Magnesium-L-Threonate (MgT). MgT has been shown to enhance memory and can even help people without an injury to maintain or improve their memory. MgT is very likely critically important in decreasing brain aging.

The dose of magnesium-L-threonate that my patients take is 2,000 mg per day.

Melatonin

I recommend melatonin to treat sleep difficulties in patients with concussions, post-concussion syndrome, acute TBI, and PTSD.

Melatonin is a natural brain hormone that helps control sleep. Our melatonin levels slowly fall with age. Melatonin production may be faulty after brain injury.

Sleep difficulties are common in patients with concussions, TBI, and PTSD. Taking melatonin may be helpful in establishing good sleep cycles to reduce fatigue and depression.

The dose of melatonin may vary between 0.3 mg and 20.0 mg at bedtime. The most common side effects of melatonin can include daytime sleepiness, headaches, dizziness, or vivid dreams. Melatonin may interact with blood thinners or diabetes medications. Check with your own personal physician before starting melatonin.

Avoid activities that require you to be alert such as driving or operating heavy machinery for 4 to 5 hours after taking melatonin. In this situation, melatonin may be far superior to pharmaceutical sleeping pills because of its relative lack of side effects.

Other Supplements

There are other supplements that may be helpful when customized to each patient.

These include:

- Vitamins B6, B12, folate and C
- Coenzyme Q 10
- Ginkgo biloba
- Acetyl- l-carnitine
- N-acetyl cysteine (NAC)
- Phosphatidylserine
- Glutathione
- Pyrroloquinoline Quinone (PQQ)

Exercise

It may seem odd to include exercise into a list of supplements. However, I want to emphasize that exercise increases blood flow throughout the body, including to the brain. Many of the supplements that I talked about improve blood flow and function. Exercise is just another tool to do this. With a concussion or TBI, exercise may be limited initially but studies have shown that slowly increasing exercise to a level that does not bring back symptoms helps healing.

Make exercise part of your daily healing routine.

NEW HOPE

Most patients in our Komer Brain Institute have post-concussion syndrome. They have been seen by neurologists, psychologists, psychiatrists, physiotherapists and occupational therapists. The reason they have been referred to me is because they are not completely better. As a matter of fact, many of the patients are only slightly improved from their initial injuries. They seem to have "hit the wall" in their recovery.

In the medical literature, I have seen 100-page protocols developed to look after post-concussion syndrome that have no mention of brain hormones at all. Some of the others suggest that brain hormones should be checked, but that is the only reference to them and there are no guidelines.

Research shows that a very large number of patients with brain injuries have hormonal abnormalities. The pituitary gland and the hypothalamus are responsible for hormone production and function.

That hormone levels change following an injury should not be a surprise. These organs are located in your head. Your head was hit. These organs are not working optimally now. The good news is that correcting the hormone levels is not rocket science! However, it is good brain science!!

The problem here is that many of the traditional professionals looking after concussions, and brain injuries are not experts in Interventional Endocrinology (the science of detecting hormone abnormalities and optimizing them).

As previously mentioned, some of these early hormone abnormalities are temporary, while new hormone problems can become apparent later. Within three months after brain injury, 56% of patients have abnormal pituitary function and abnormal hormone levels. At one year, 36% of patients still persist with poor hormone levels.[12] [13]

When I have lectured to professionals and presented these findings, most are very surprised. Many are happy to learn that there is an area where the patients may be helped. However, many physicians looking after brain injuries have no specific training in these areas. One of my goals is to use lectures and other educational tools to update all professionals about the need for optimal hormone levels for optimum brain chemistry, function, and for healthier happier patients.

The other important new insight is that the brain becomes inflamed by an injury. In retrospect, this should come as no surprise. Everyone realizes that if you break a finger, there is going to be a process of swelling and inflammation. However, this inflammation usually slowly heals on its own. The brain is different, and the inflammation can continue for years and decades after the initial injury.

These smoldering coals of inflammation, if unchecked, inevitably lead to the damaging flames that cause cognitive, emotional and neural degeneration.

It may well be that this inflammation makes the brain more vulnerable to greater damage with each successive injury.

What does this mean to someone with concussion or TBI?

---

12 Zhang L, Yuan Y, Tong Q, et al. Reduced plasma taurine level in Parkinson's disease: association with motor severity and levodopa treatment. Int J Neurosci. 2015 May 23:1-24

13 Journal of Endocrinology Investigation 2005:28 Popovich et al

If you have not had a thorough hormonal workup by an expert, then your testing is not complete.

If you have hormone abnormalities (and remember only optimal levels are acceptable), then you may still have new treatments, which can improve your wellness considerably.

Let's fix those hormones.

If you have not been treated specifically for brain inflammation, your treatment has not been complete. Reducing the brain inflammation improves the toxic environment in which the brain currently exists. When this happens, the brain has a much greater ability to heal itself (brain plasticity). You can still improve.

So what is the new hope? We have found two major problems in your brain: abnormal hormone levels and inflammation. If these have not been addressed yet, they can be treated. They are fixable.

There is new hope.

MENTAL GYMNASTICS

Many choices contribute to our health over a lifetime. We tend to think about our heart health, exercising, being mindful of what we eat, and sleeping well. All of these decisions made day in and day out are good for our body. Our brain health also benefits from our overall wellness.

Just as we target certain muscle groups with specific exercises designed to stretch and strengthen them, there are activities that will help our brains stay healthy too. When you have suffered a brain trauma, the importance of healthy choices is even more vital.

We will discuss the healthy choices of nutrition and supplementation elsewhere in the book. Here, we will consider activities that can add to your brain health and your brain power.

Being a lifelong learner is an important concept to keep our employment skills sharp in an ever-changing economy. To compete, we must continually build on what we know and discover new ways of doing our work. The lifelong learner principle not only helps our career, it helps our brain cope with the challenges of aging as well. Learning keeps our brain engaged in ways that stimulates our brain to be strong. Doing activities that require creativity, and curiosity help us expand our thinking and our horizons. You might even have some fun!

What will help?

Here are just some of the myriad possibilities that you can choose to do.

Play games. The challenge of playing a game will engage your mind to solve problems and respond to something new. Enjoy playing the games you know. Keep up your skills in playing cards. Work away at the crossword puzzle or Sudoku challenge. Challenge yourself on a new app that will require you to learn how to play the game and then overcome obstacles to succeed. Get out that old board game and enjoy a fun time with family or friends. It is helpful to play different games that involve a variety of learning skills. Do not just do math games or only word based challenges. For those who prefer a system to help you do it, search out the many apps that are available. Lumosity has a website and mobile apps with programs designed to keep your mental agility high. Mix it up and your brain will thank you.

Music is a powerful friend for our brains. Learning to play an instrument is like learning a new language with the bonus of connecting our brains to the parts of our body used in playing the instrument. The rhythms, harmonies and melodies of singing engage our brains and stimulate our neural pathways. Even listening to music keeps our brains functioning at a higher level of activity as we interpret the sounds and expressions we hear. Music can also help us relax and be mindful.

Read something. Never have we had so many ways to read and learn. Books were once the domain of the wealthy and the educated few. Now we can source books over the ages and in many genres easily. Just as different kinds of music may stimulate our cognitive powers in varying ways, reading a variety of books is helpful. Fiction, news, history, biographies, science, and nature all benefit our brain to think about something new or engage ideas in a new way.

Build your relationships. Spending time socializing with family and friends provides emotional support through the ups and downs of life. By having conversations, our brain is required to deal new information that someone might share with us. Questions give us an opportunity to respond to or memories or ideas. Social settings also challenge us to adapt and interact with our environment as we relate to other individuals or to the group. As you soon learn, dealing with relationships is complex. While we might sometimes wish that they were simpler, at least we have the comfort that our brain is being challenged even as these dear people drive us crazy! Relationships are part of our well-being. As therapist, Virginia Satir said, "We need 4 hugs a day for survival. We need 8 hugs a day for maintenance. We need 12 hugs a day for growth." Human interaction and social support networks are especially important when you are ill. The more support you have, the better your body and mind can recover.

Start a hobby or develop an interest that is not related to your work or daily routines. Become a gardener. Learn how to build something. Discover the joy of painting. Take cooking classes. There are so many activities we can do that will require our brains to learn, adapt and create. Do not put these hobbies off until you retire. They are beneficial when you are still busy in your working years. Your retirement will be happier and healthier when you enter it with interests already established. Retirement then becomes the opportunity to do even more of what you love as well as the time to add other skills and hobbies too. However, if you are already retired, take the plunge into something new.

Play sports to keep your brain strong. Participating in sports creates mental challenges as well as adding to your physical well being. Although you may not be able to continue the sport you loved when you were younger, it is great to take up a new sport that will add to your life. Choose activities that are appropriate to your particular age and stage of life as discussed with your health professional. Many sports now have modified versions that make it easier for people who are aging or have other participation challenges. Start where you can participate comfortably and progress from there. There is no shortage of sports and other activities for every age and stage of your life.

Serve others. Find an opportunity to care for others in your community. As you make a contribution to the lives of other people, you engage in situations that require you to think in different ways than your work might require. Any activity that includes socializing and learning something new also adds to your mental strength. Connecting to community will also give you a type of satisfaction that is different than doing your job. All of this helps your sense of well-being.

Think of your brain as you do your muscles. It needs exercise and challenges to develop and to stay strong. Where you have the added challenge of a brain injury, choosing to keep your mind learning and growing will help your recovery too.

FORGIVE AND FORGET

When someone experiences an injury that alters their lives, one of the challenges that are often overlooked is forgiveness. It is natural for us to try to connect a cause and effect for what happens in life. We usually go beyond that by looking for whom to blame for what has happened.

Blame the driver of the other car who crashed into me or the football coach who told me to hit them hard. It was the fault of the commander who ordered us out on that patrol. Fault the store owner who left an icy patch on the sidewalk where I fell and hit my head. All the many ways people experience traumatic brain injuries, concussions, and battlefield trauma usually have a "somebody" to blame for what happened. Even when there is no one, in particular, we can point to – we blame God or the Universe for our loss.

Whoever else there may be to blame, we also usually include ourselves in the list of guilty parties. This is especially true when an injury is the fault of our carelessness, risky behavior, or stupidity. However, even when we were not at fault, we ask why we took that street at that time, why we signed up for the hockey team, or why we joined the military.

There is usually a whole lot of blame to go around. Blame leads to unresolved pain and bitterness. It will rob you of peace and even your health. In a brain injury, the scars are not just physical but psychological, emotional, and spiritual.

Whatever has happened in your injury may involve legal or other considerations that should be pursued. This chapter is not about that.

This is also not about the moral responsibility or social policy surrounding the epidemic of brain injuries.

Something terrible has happened to you. What can you choose to do to promote your healing and well-being?

The power to forgive is a great gift we have all been given but something we seldom use. It can be part of our recovery too.

One of the common misconceptions when we talk about forgiveness is that it is better not to forgive: "That person does not deserve to be forgiven" or "What they did to me was so painful or hurts so much that I don't want to forgive them. I refuse."

That misses the point of the gift we have received with forgiveness. Forgiveness is primarily for the benefit of the person doing the forgiving.

Yes, there are certainly some benefits for someone who has been forgiven, and that is important, too. Even so, the first and principal benefit comes to the person who is willing to forgive. That's why we call it the power to forgive: When you choose to forgive, you are actually exercising your power in that relationship to move it forward.

Forgiveness is about how we process our feelings on the personal side.

We tend to invest a great deal of energy when we blame others or ourselves for something that has happened. Our anger at their careless or willful acts consumes us. Where we are to blame, our self-loathing, because of something we did to cause our accident will risk our emotional health.

Guilt and resentment have serious health consequences. Not forgiving yourself, or others, keeps you in a state of negativity. That negativity weakens the immune system at a time when a healthy immune system is is so necessary to recover.

As the old saying goes, "Not forgiving someone is like taking poison and expecting the other person to die"!

So what can we do?

We can choose to forgive that person and ourselves for what has happened. That has nothing to do with the legal or moral responsibility, but it is a decision we can make in our hearts for our future.

By saying, "I am choosing to forgive that person." does not deny your pain, or claim that what happened did not happen. It does not assume that they deserved forgiveness, asked to be forgiven, or are even available to meet with you. In legal proceedings, you may be required not to meet with them.

You may be correct to take responsibility yourself for a decision you made that caused your injury, if that is the case.

However, by making the decision to forgive, you choose to set yourself free.

Forgiveness is not about something that is deserved – it is a gift given. Like most things in life, when we are generous, we receive much more than we gave away. The same is true when you forgive someone else. You give them the gift of moving on. You give yourself the gift of not being chained to them emotionally.

Often, the most difficult person to forgive is you. If we do not choose to forgive ourselves, we will be trapped within a circle of regret and self-pity that will only add to our burden. It will make our recovery that much more difficult to achieve. Bitter people tend to be corrosive in their relationships. They damage the very people closest to them, the very ones who can support and encourage them during their healing journey.

Even beyond forgiving, it is possible to forget, as well. That may sound impossible and you may be tempted to say, "I will forgive, but I won't forget!" That also keeps you from the more complete experience of moving past those things that hold us back.

The next time you are reminded of their role in your injury - practice forgiveness. Say to yourself, "Yes, I choose to forgive that person." By doing that, you choose to move on, and that is very liberating. We encourage you to get into that practice of forgiving.

Once you develop the habit and practice of choosing to forgive, you will be surprised to find out that you really can also forget as well. That gives you the freedom to break the chains and move on to become your best self. This is true not only about the person involved in your injury; it is true in all of our relationships.

Do not forget to include yourself in the list of people to forgive. We all have moments big and small where we wish we made other choices. Some of those decisions have huge consequences. The good news is - you are human. The bad news is – you are human. Our humanity inconveniently does not include perfect judgment. We will make mistakes. Forgiveness is how we move forward when our choices hurt others or ourselves.

Choosing to forgive will not turn back the clock to life before your injury. It will ensure that you do not spend all of your future focused on being angry at that other person or yourself. You deserve better. Forgive and you will find it.

AN UNEXPECTED COMPANION

Waiting for all of us on the journey after any loss is an unexpected companion. This traveler will join us for a long time to come. It is named Grief.

It may be a lost relationship through death, divorce, distance, or disputes. Financial losses or employment disappointments can impact our life dramatically as well. Even a shattered dream can send us into a tailspin.

Here, what we will consider is a loss of health. Injuries rob us of our present contentment and often our tomorrows too. Suddenly, we cannot do all the things we once did. Perhaps we cannot play the sports we loved so much. It might be more difficult to do our work. We may not be able to do the simplest of routines that were so easy just a short time ago. Before long, we look at our present condition and imagine our life in our future. We begin to wonder how we will ever cope with this changed life.

When the health loss involves our brain, the impact is even more profound. Not only might we have new physical limitations, but we may have new challenges in how we think and feel. That is why brain injuries, battlefield traumas, and concussions cause so many additional barriers to recovery. We are not just trying to rehabilitate an arm or a leg – as difficult as that process is. When your brain has been injured, your mind, will, and emotions may also be compromised.

From the moment we experience a loss; we begin a journey through life after that event. Some aspects of the loss may be recovered over time while others may never return to what once was. Losses mean change. All change brings work to be done for us as persons. Part of that work always includes Grief.

If you have experienced significant losses in your past, you may already be well acquainted with this traveling companion. For those who have not walked the road of loss before, there will be many lessons to learn along the way.

It may come as a surprise to think of Grief as a gift. We are given Grief as a resource to help us endure our losses and to find a way forward. How we respond to our friend Grief will determine how our future will unfold in many important ways.

Swiss-American Psychiatrist Elisabeth Kübler-Ross identified these stages: denial; anger; bargaining; depression, and finally, acceptance. You can read more about this In her 1969 book, "On Death and Dying." The stages are there. You will go through them as you experience the grieving process. There are no shortcuts. None of the stages of Grief can be skipped. You also cannot rush the experience - no sprinting allowed. You may postpone the journey. Grief will wait patiently for you until you are willing to start the long road together.

If you imagine meeting Grief after your loss in the beginning of your journey of recovery, the first response you would have is to ignore this would-be companion. The stage of denial is a defense mechanism that, like going into shock, freezes us in the moment. It will take a while to understand what has just happened to us. We cannot believe it. This cannot be true. It is only a bad dream. Somehow, we will wake up and everything will be back to what it used to be. Strangely, as we are trying to ignore our loss, we are actually beginning our recovery. Denial is the first stage of the grieving process. Everyone will go through it, sooner or later.

The next stage is to be ANGRY! We yell and scream in our mind heart – and sometimes out loud too. Imagine Grief listening to us vent our rage at how unfair this loss is. We have been wronged by that other person, by life, or by our own choices. Our anger wells up from deep places in our being that might surprise and even shock us. However, Grief refuses to abandon us, and our journey continues.

The stage of bargaining is next. Here we experience our personal version of "Let's Make a Deal!" Negotiations begin. We suggest that in the future, we will do something important or sacrificial in return for our loss being made whole again. Our appeal to God or the Universe offers an unyielding commitment - if only we can go back. No matter how persuasive we are, the loss continues to be real. We move past the stage of bargaining as we journey further.

Grief then leads us to meet our next stage called depression. Frustrated that our bargaining did not succeed, our anger turns inward to a sense of helplessness then hopelessness. Down we walk into that dark valley where many people wander for a very long time. Identity and life become defined by the loss experienced. Grief waits for us while we are in that hopeless place, beckoning to take us on to the final stage of recovery.

Our final destination is the land of acceptance. We come to understand our loss in a different way. No longer do we deny it, even though we wish it never had happened. Finally, we recognize that there was no deal to be made to turn back the clock. The depression lifts as we move forward into that land of changed persons. We begin to inhabit this new place where we define ourselves not only by our achievements and gains, but by our losses too. A more mature person has emerged from this dreadful experience.

The memory of our journey with Grief will always be with us. What we have endured ultimately will free us to go on in spite of it all.

MINDING YOUR MOMENTS

Mindfulness is a word that has become popular recently, but its origins date back thousands of years. This idea of mindfulness is about paying attention to the moment you are experiencing.

In Western societies, we live life with our shoulders forward leaning into the future. This has many benefits as it helps us to strive for something better, and to imagine a brighter tomorrow. Innovation and industry rely on this vision of what might be. In the information age, we are bombarded every moment with news from around the world delivered by television, the radio, in your email alerts, through your news feed or, with a notification on your mobile app. This wave of information washes over our moments, influencing our understanding of life.

What we have lost in this worldview is the ability to take time to consider what is happening right now inside ourselves. This was an assumed part of many ancient cultures and religions. The benefits of meditating, prayer, reflection, and finding calm are not new ideas. Theirs was a practice to take time in a quiet place, to understand your life, and once again your mind and heart.

One of the pioneers in rediscovering mindfulness in our time is Harvard psychologist Dr. Ellen J. Langer. Her work on this topic began many years ago. She has been very influential in deepening our understanding of the nature and value of mindfulness. Her book "Mindfulness" was re-released in 2014 in a 25th anniversary edition.

Mindfulness includes a number of different challenges. Perhaps the most difficult is the choice to let go. We spend so much of our life trying to be in charge of what is around us (including our future!) that giving up control is a hard concept to put into practice.

What happens when you practice calming yourself and quietly reflect? You begin to become aware of what is going on around you, and inside of you in a new and deeper way. You begin to hear the still small voice deep inside your soul that has been drowned out by the noise of a busy life. Even your awareness of your body begins to change. As you breathe, you begin to sense the many things that are happening inside. Like being still in a forest, new sounds are heard that you missed when you were busy trampling the undergrowth. By letting go, a new freedom to experience the present occurs. No longer exploring the future, you can discover yourself and the moment in time that we call "now."

Many benefits take place when we take time to mediate, pray or reflect. The calming of our mind brings a peace over our emotions and body. The long list of fears no longer seems as pressing. Stress and negativity do not only rob you of your sense of well-being; it can also compromise your health. By giving up control in favor of quietness, you loosen the grip of those negative emotions that seek to control your every moment.

The practice of mindfulness does not come as easily to some as to others. Those personalities who are driven and value performance over everything else will find it is a challenge to relax and reflect. It is as difficult for them to slow down as it is for an introvert to face a crowd. The keyword here is the "practice" of mindfulness. It certainly does not seem natural in the beginning. However, the practice will help you extend the experience further each time until it becomes a normal and comfortable experience.

This resting of your mind is different than when we fall asleep. In the many stages of sleeping, we can be very active mentally as we dream.

Our dreaming mind is beneficially working through the events of the day, the concerns of tomorrow, along with the many strange encounters we have in the dream world each night. So strangely enough, our ability to choose to be calm and mindful is a key to finding that deeper level of peace.

Mindfulness is not denying the world around you. It does not discount the real concerns you may have about our family, friends, work, finances and health. However, by being mindful, you move away from all of that to a place in your soul where you can meditate, pray, or reflect upon your life. You find a safe place – a sanctuary – from the cares of the world. Richard J. Foster wrote a book entitled, "Sanctuary of the Soul: Journey into Meditative Prayer" that discusses the practice and value of this journey.

Some health conditions make the practice of mindfulness a challenge. Those who are clinically depressed, anxious, or have obsessive-compulsive disorder find it difficult to live in the moment. Finding professional help with these conditions may help you discover mindfulness.

In the Alcoholics Anonymous recovery program, the participants recite The Serenity Prayer. In asking for wisdom, there is a release of control (In both AA and Al Anon, this is an appeal to the Higher Power for help.) The prayer talks about the wisdom to know the difference between what we can and cannot change. We must change what we can, not live in fear, and accept what we are unable to change.

For those who have experienced a brain injury, this practice can aid your recovery. For others, it is a way to renew yourself, to free yourself from the burdens of the day. This unlocks our creativity and sensitivity to those in our lives.

Explore the value of practicing mindfulness. Discover the moment.

LEGAL CHALLENGES

Many cases of concussion, TBI and PTSD involve legal action. I am seeing several patients who pursued legal action and still are not well. These patients may have had a slip and fall, motor vehicle accidents, workplace injuries or injuries from sports.

This damage can happen because of equipment failure. It might be due to no fault of the victim. Still, you may experience a serious injury. The patient may be entitled to benefits or may have settled the claim. Later, the full extent of the injury is apparent.

This is particularly true of the hormonal changes that I see. They may appear days, weeks, months, years, or decades after the initial injury. The patient, most physicians, and most lawyers remain unaware of the extent of hormonal damage.

The relatively quick onset of symptoms such as headache, head pressure, dizziness, nausea, sleep disturbance, and light sensitivity are quickly diagnosed as post-concussion symptoms. However, the onset of post-concussion symptoms such as fatigue, depression, reading difficulties, difficulties with comprehension and memory, decreased competitive drive, loss of consciousness and sexual changes may not have been attributed to the original injury.

Similarly, with PTSD, the changes are often said to be caused by stress, but the hormones may not have been evaluated. Particularly troublesome are long-term serious problems. Low estrogen levels in women or low testosterone in men increase a number of risk factors. It can lead to an increased risk of stroke, heart attack, diabetes, depression, and arthritis.

In short, death rates may rise, due to poor hormonal function.

Often patients get to a point where they are told that nothing further can be done. They may be told that there will be no further improvement. Taking away hope itself, may lead to depression.

I will never tell patients that nothing more can be done. If hormones have not been evaluated, there is hope for considerable improvement.

The other thing I see is patients being told that everything is in their head. There is nothing wrong. They should be able to do everything. They should go back to work.

What I want to make very clear is that every assessment of concussions, or PTSD should include all of the following:

- Hormonal and nutritional evaluation
- Assessment of visual processing
- Assessment of hearing and balance
- Assessment by an expert in interventional pain management of whiplash and other muscular or joint injuries.

Patients are often sent for neurocognitive behavior therapy. This is to assist them with coping strategies with the remaining problems, as opposed to treating them. I think this training has its place, but not until all of the above evaluations have been carried out, and appropriate treatment given.

Often sufferers are asked to settle their claims prematurely. Even if the immediate problems are stable, we know that further long-term difficulties can occur later, and this must be taken into account before any settlement is reached.

Our clinic is a resource for patients, and their legal teams, we are committed to ensure that deserving patients will have a more complete assessment, and treatment, as part of their settlement.

Our goal is always to achieve as complete a recovery, as is possible. New techniques and ideas make this more and more likely every year.

ANTI-AGING STRATEGIES FOR YOUR BRAIN

Hormones are important throughout our lifetime. When we are young, they help us grow and mature into adults. They are partially responsible for how we look, how we act, and how we feel. In many ways, they also keep us healthy. We have already discussed the tremendous effect of restoring and maintaining healthy hormones, after a brain injury, or PTSD.

During our lifetime, there are three major health issues related to aging. These are heart disease and stroke, cancer, and brain aging, also known as neurodegeneration.

We have come a long way in defining the risk factors for heart disease. There are lifestyle choices we can make to reduce these risks. We are much better at preventing heart disease by treating high blood pressure, high cholesterol levels, and diabetes. There has been a tremendous push to eliminate smoking, which will lower the risk of heart attack. We are better at treating heart attacks to minimize death rates. Better cardiac rehab has led to further improvements.

As far as cancer goes, we have found ways of preventing it. We know that decreasing smoking is a preventive measure to reduce lung cancer. We have learned that reducing the spread of human papilloma virus (HPV) can diminish gynecologic cancers, particularly cervical cancer, as well as reduce the risk of other cancers, such as oral, and anal cancer. We now have vaccines that reduce HPV infection, and lower the risk of certain cancer. Exercising more, not gaining weight, and drinking less alcohol can dramatically reduce breast cancer. Earlier cancer detection leads to a higher cure rate. Our success in eradicating cancer with surgery, chemotherapy, and radiotherapy improve every year.

The next big hurdle to maintain health, and minimize the effects of aging, is to reduce brain degeneration. Rates of Alzheimer's disease and other dementias, as well as Parkinson's disease, appear to be increasing. Almost everyone fears that his or her brain will degenerate with age.

We know that traumatic brain injury raises the risk of Parkinson's disease, later in life. Brain trauma can lead to chronic traumatic encephalopathy (CTE). The movie "Concussion" has quickly spread the news that athletes are more prone to CTE. This is a form of neurodegeneration that occurs earlier in life than other dementias.

What goes wrong in the brains of young people that cause a condition like CTE?

Why do some players get it in others do not?

How can we prevent it?

Athletes in collision sports have a much higher rate of concussions, and traumatic brain injury than athletes in none collision sports and in the general population. Brain injuries appear to be a risk factor for CTE. This appears to be a risk factor for CTE. These same individuals form one of the highest risk groups we see in the Masters Men's Clinic. We have seen athletes as young as 18, who have extremely poor hormone levels. We know that the restoration of these levels improves their function. We do not yet know if it slows the degenerative process. However, science suggests that this may be the case.

Here are some other important facts.

Almost all aging is a combination of inflammation, reduced blood supply, and in some cases, the accumulation of toxic substances.

For example, in Alzheimer's disease, there is the development of "plaques" which are numerous, tiny, dense deposits scattered throughout the brain that becomes toxic to brain cells at excessive levels. These "tangles" interfere with vital processes, eventually choking off living cells. When brain cells degenerate, the brain markedly shrinks in some regions.

"Plaques" form when protein pieces, called beta amyloid, are clumped together. Beta-amyloid comes from a larger protein found in the fatty membrane surrounding nerve cells. Beta-amyloid is chemically "sticky," and can gradually build up into plaques. The plaques may block cell-to-cell signaling. They may also activate immune-system cells, which trigger inflammation, and devour disabled cells.

"Tangles" destroy a vital cell transport system made of proteins. The transport system is organized in orderly parallel strands, somewhat like railroad tracks. Food molecules, cell parts, and other key materials travel along these tracks. Protein called "tau protein" help keep the tracks straight. In areas where tangles are forming, tau protein collapses into twisted strands, called "tangles." The tracks can no longer stay straight. They fall apart and disintegrate. Nutrients and other essential supplies can no longer move through the cells, which eventually die.

Plaques and tangles tend to spread through the brain into the predictable patterns in Alzheimer's Disease. The early changes of Alzheimer's disease may begin 20 years before diagnosis is made. Mild to moderate Alzheimer's disease generally lasts 2 to 10 years. Advanced Alzheimer's disease may last from one to five years, and is fatal.

The latest studies show that the tau protein accumulation leads to cell death, and is the main problem with Alzheimer's disease. Inflammation is also part of the process.

Our original question was, "Can this degeneration be prevented?"

One of the problems with brain degeneration is decreasing blood supply. We know that certain things increase blood supply.

One of them is exercise. Therefore, lifelong exercise is critical.

We know that for women in menopause, blood flow to the brain drops 30%. Restoration of estrogen restores blood supply to the brain. There are studies that show that estrogen replacement also reduces Alzheimer's disease. This is not surprising, since one of the other properties of estrogen is that it is a good anti-inflammatory in the brain.

In the clinic, we have also learned that similar to estrogen, testosterone in men is a good vasodilator (increases blood supply.) It is also a very strong anti-inflammatory. From a scientific point of view, it would seem that restoration of these hormones for a lifetime is critical in preventing neurodegeneration.

Another important fact is that both Vitamin D, and Omega-3 fish oils in high doses, are potent brain anti-inflammatories. I use both of these in the immediate treatment of concussion, and for post-concussion syndrome, since it is felt that part of the concussion process is brain inflammation.

We know that the PDE-5 inhibitors such as Cialis, Viagra, and Levitra increase blood flow, not only to the penis, but also to many other areas within the body, including the heart and brain. Spasm, or closing of blood vessels, occurs following concussions, and leads to decreased oxygen supply. Can these potent drugs help both the acute concussion, and post-concussion syndrome? Will these drugs be helpful for maintaining healthy brain function for a lifetime? We have already seen indications that low-dose daily Cialis may reduce heart attack and stroke, as well as reduce prostate cancer rates. In time, we may see proof that these drugs are helpful in concussion and TBI and may also slow neurodegeneration.

There is exciting new information suggesting that a highly absorbable form of magnesium called magnesium-L-threonate, crosses the blood-brain barrier and may reverse neurodegeneration. This substance seems to be very helpful in improving memory, which is often a feature of concussion, and TBI. We will talk more about it in the chapter Brain Food.

It appears that many of the treatments we use for post-concussion symptoms could work in healthy brains to continue to keep them vital throughout a lifetime.

We are learning how existing medications might be applied in new and different ways to combat neurodegeneration. The same substances which are helpful with TBI, and post-concussion syndrome, might also be beneficial in preventing or treating Alzheimer's disease.

There indeed is New Hope!

IT TAKES A TEAM

Concussions, traumatic brain injuries and PTSD are incredibly complicated conditions. I spent 40 years, after I graduated as a specialist, learning about hormones. We know that medical knowledge doubles every 11 hours. It is difficult to keep up to date. During these years, I have dedicated one or two hours a day researching hormones and applying this knowledge, to diagnose and treat my patients. In the earlier years, this meant reading scientific journals, talking to other interested professionals, and attending healthcare and research meetings. Since the advent of the Internet, learning is so much easier. Information is disseminated quickly.

However, during all this time, the most important learning has been from dedicated colleagues both in my field, and other areas of medicine. Sitting down and talking to such colleagues for an hour is invaluable. Hearing about their results from treating patients is more helpful than attending two or three-day seminars, or even reading countless journals.

We not only discussed what worked to help our patients but, more importantly, we spent many hours discussing what did not work. This is valuable knowledge so that time is not wasted in the future with techniques or treatments that are not helpful.

The most brilliant people are those who can think outside of the box. They are constantly devouring new information as they figure out unique ways to apply this knowledge to help patients. I have seen too many academics, who have a great knowledge of research, but little understanding about how to apply this to help patients get better.

I have treated hormone problems in women for 40 years. I have learned from constant feedback from 13,000 menopausal patients what works effectively. I hear daily how the restoration of hormones has been life-changing or even life-saving.

Some 15 years ago, after listening to many female patients ask what I could do for their husbands, I had to take a fresh look at what I thought I knew. They often told me that their husbands had the same menopausal condition as they had. I carefully explained that no such problem existed. However, when you have several hundred people telling you the same thing, you have to look at your information and see if it is faulty. That is exactly what happened to me. They were right. I was wrong.

As I studied men's hormones, I realized that Andropause (or Low Testosterone) led to symptoms and chronic illness, just like menopause. We have seen more than 6000 men in the Masters Men's Clinic dedicated to diagnosing hormone deficiencies and optimizing their levels. This resulted in men feeling so much better, and also in a reduction of serious illness, such as stroke, heart attack, and depression.

During all of this time, I was involved in Sports Medicine. My original degree was in Exercise Physiology, which is the science of exercise and sport. I played hockey for 50 seasons as well as played lacrosse. When I was younger, I coached, and was the team physician for many amateur sports teams. Often, they were the teams on which our son, Scott, played. These were the hockey and lacrosse teams with which he started as a three-year-old. After many years, and many teams, he eventually became an accomplished professional lacrosse goalie, winning Canadian and international championships, and awards.

I saw concussions along the way. Then, I started seeing a pattern where concussion symptoms were almost the same as the symptoms of low estrogen in women, or the symptoms of low testosterone in men.

This led me to start treating post-concussion syndrome and PTSD. Much of the time, I diagnosed low hormone levels, and patients got better. Now that I have had the opportunity to look back on the 6,000 men we have treated in the Masters Men's Clinic, it appears 90% of them have had head injuries. Many of the injuries occurred years, or decades before they developed low testosterone.

Before long, I realized that there was much about concussions and PTSD that I did not know. I was determined to understand what was going on when these problems occurred. I found that other specialists were very important in treating brain injuries and PTSD. I have had the incredible fortune to meet talented and passionate individuals in several fields that treat difficulties after concussions. Combining their understanding of post-concussion challenges with my understanding of hormones, the insight and protocols were developed. Together, we have had much greater success in helping patients than I ever could have had on my own.

Clearly, no single person alone can successfully treat concussions, TBI, and PTSD. It takes a team. Just as it is in sports, every member on the team is important and necessary.

My very good friend and colleague, Dr. Frank Stechey stated: "It is amazing what a group can get done, when no one person has to take the credit."

That is the way it is with concussions, TBI, and PTSD.

I am very proud of my team and want to tell you about them.

Dr. David Levy is the premier Sports Medicine physician in Canada. He started looking after athletes before Sports Medicine became a specialty. He has trained countless Sports Medicine physicians, who have become leaders in Sports Medicine. He has shared his knowledge with many other professionals such as athletic therapists and trainers, who have gone on to distinguished careers. I have seen him treat elite professional athletes and children just starting their sports careers with the same kindness, compassion, and excellence. He has looked after my family and me when we have had sports injuries. I have had the honor and privilege of working with him with lacrosse teams, and in some other sports.

He has been Team Physician for the Hamilton Tiger Cats Professional Football Team for more than four decades. We have worked together as Team Physicians for the Toronto Rock Professional Lacrosse Team since its inception, 18 years ago.

His guidance, encouragement, and friendship mean the world to me. He is always kind enough to give his expert opinion and treatment of athletic injuries, to my concussed patients.

Dr. Patrick Quaid is a brilliant Doctor of Optometry. He routinely sees many of my patients with concussions, TBI, or PTSD. It seems that almost every patient with these problems also has difficulties with visual processing. Dr. Quaid has taught me that there are 15 features of vision; one of them is assessed by the eye chart on the wall, but there are 14 others that often are not checked. Our patients frequently have symptoms of headache or difficulties with reading or light sensitivity. Visual processing is incredibly common with brain injuries. Patrick has been kind enough to write a chapter in this book, and you will read of the importance of what he does in getting our patients better.

Recently, Dr. Marisa Marchionda, another gifted Doctor of Optometry, has been added to my trusted team and is also assessing and treating eye problems and visual processing difficulties that occur in concussions, TBI and PTSD.

Dr. Blair Lamb is a gifted physician who has dedicated his medical practice to the management of pain, as well as to the treatment of muscle and other problems that occur with concussions and brain injuries. It seems that almost 100% of people with a brain injury also have a neck injury or some other muscle problem accompanying it. Blair has developed his own protocols for treating these patients. I can honestly say that there is not one patient who I sent him who was not substantially improved after his care. His aim is to cure the injury, and not just to reduce pain, or have patients need a lifetime of treatments. Blair has been kind enough to write a chapter in this book.

Dr. Brenda Berge is a Doctor of Audiology. Audiology is the branch of science that deals with hearing, balance, and related disorders. Many concussed, and brain-injured patients have problems in this area. Brenda has been a great help in diagnosing and treating our patients with hearing issues, such as noise sensitivity, tinnitus (ringing in the ears), balance issues and dizziness. She often sorts out which of these problems are due to the ear, and which are due to visual processing problems.

Dr. Frank Stechey is a close friend and an expert in sports dentistry. He is the team dentist to the Toronto Rock Professional Lacrosse Team and has consulted throughout his professional career with athletes in many other sports. He is a noted international speaker on dental sports injuries. His expert care, particularly in fitting customized mouth guards, helps prevent concussions, and reduce facial injuries.

Dr. Terry Moore is a physiologist in Guelph, Ontario. He has looked after the muscular injuries of a large number of patients referred to me. He is dedicated to improving the lives of patients with these injuries, and utilizes some innovative treatments. He is an expert, in his area, as it relates to concussions.

Joan Komer is a special member on the team. With her degree in Psychology, and with Masters of Education, she is a motivational speaker, and my wife. I have heard her give more than 100 interactive seminars to large audiences, and her message of hope and being "the best that you can be," has affected the lives of many, including me. I carry her message into my practice, not only treating patients, but assuring them that there is always more that can be done to help them improve. She has continually encouraged me in my career, especially when I have embarked on new paths, such as treating concussions, TBI, and PTSD.

In my office, nurse Heather Lawson has been a pleasant, valuable contributor to the daily care of our patients. Sarah Hepworth manages the office and keeps things running smoothly, always with a smile. Gordon Tonnelly is the Director of the Masters Men's Clinic and is a skilled counselor and researcher.

So, that is our team. They are brilliant professionals who are passionate about helping people. Each is a creative thinker who is on the leading edge of knowledge. As practical experts, they use their considerable experience combined, with the latest science, to help patients with brain injuries, and PTSD.

I am incredibly grateful to have every one of them in my life.

Each one, in his or her own way, is advancing the science of treating concussions, TBI and PTSD.

TRY TO REMEMBER

William S. Cook, Jr., M.D.
Board Certified Psychiatrist

As we all age, there will be noted changes in "cognition" or brain functioning. Most of us will personally encounter some type of dementia, or have a loved one or friend who will develop it.

Normal brain functions in different areas within the brain include: learning, memory, language, perceptual/motor skills, complex attention, executive functioning, and social cognition. Over time, it is common to have some progressive forgetfulness. Perhaps we forgot where we placed our keys or our purse. Maybe we forgot a doctor's appointment or a loved one's birthday. It is normal to have some forgetfulness, as we get older. There are those for whom forgetfulness increases faster than for others.

What is not normal in the aging individual is a medical illness of the brain called dementia. Dementia is a broad term for all brain illnesses that affect memory, concentration, language skills, and executive function. Everyone will have some level of forgetfulness, but not everyone will develop dementia. Dementia is usually diagnosed where there is a gradual decline in cognitive functioning. There is a noticeable impairment of independence in everyday activities. These activities may be as simple as managing medications, or paying bills.

The most well-known form is Alzheimer's type dementia. This type of dementia occurs in approximately 50% of patients with dementia. It is difficult to distinguish it clinically from other types of dementia. On autopsy, though, the microscopic impressions of Alzheimer's type dementia show a very specific quality that can be observed.

The second most common type of dementia is a vascular type dementia that occurs as a result of strokes or hardening of the arteries. Plaque build-up in the blood vessels of the brain leads to decreased blood flow, resulting in damage in the brain and the clinical symptoms of dementia. There are many other types of dementia, but Alzheimer's dementia and vascular dementia are the most common.

How does one know if they, a friend or family member has dementia? It is a clinical diagnosis that requires a history of functioning, either documented on standardized neuropsychological testing, or other clinical assessments, or documented by the decline of functioning known to close friends, or family members. The immediate and short-term memory of individuals living with dementia is more obviously affected in earlier stages of the disease. Late-stage dementia affects your long-term memory. This would include the ability to recognize familiar faces.

Individuals with dementia can often remember very clearly what they were doing 50 years ago, but cannot remember if they had breakfast, or not or whether they are taking any medications. Short-term and long-term memories are laid down in the brain very differently.

Knowing what to do if you, a family member or friend develops dementia can be very helpful. Widely prescribed medications that can affect the progression of the illness, but these will not cure dementia. It is noted that individuals with certain types of dementia respond better with these types of medications in the earlier stages of the illness. Later stages of the illness appear to be less responsive to currently available medicines.

In later stages of the illness, a small percentage of individuals might have some significant behavioral changes that might require antipsychotic medications or mood stabilizers. Geriatric psychiatry units are available to assess and treat individuals at different stages of dementia and with distinct mood disorders, and behavioral reactions that occur as a result of the dementia.

A common mood disorder associated with dementia is depression. When patients start realizing that they are unable to do the things that they could do when they were younger, clinical depression may ensue. Support of close friends and family members is essential when depression occurs with dementia. It is also possible that some type of prescribed antidepressants could minimize the symptoms.

Taking care of an individual with dementia can be extremely exhausting. While you may have great intentions to care for your parents, friend, or spouse who develops dementia, that may not be enough. Dementia can be a very long-term and debilitating process. No one person can take care of an individual with dementia. It takes a group of people either, on a dementia unit in a nursing-home setting, or group of people in a home environment.

There is a new trend across the United States and Canada for the development of what are called "greenhouses." These are places to live for individuals with dementia that provide a cheery and relaxing environment along with needed medical care. The layout of these greenhouses allows for family members to visit and participate in meals. Everyone has their individual room, which opens up into a large gathering area for the meals and socialization. These greenhouses do not have the same stigma attached to hospitals and traditional nursing homes.

Whether it is just forgetting where your keys are, forgetting large blocks of life, losing the ability to remember things, or function in social settings, everyone will experience some level of forgetfulness. It is natural for individuals to want to be as independent as possible, but it is also important that, when individuals become a danger to themselves, or other people, activities such as driving are curtailed.

Educate yourself as to the signs and symptoms of dementia so that you, a family member, or friend can be better prepared to deal with what may lie ahead.

SPINAL CONCUSSION SYNDROME

G. Blair Lamb MD, FCFP
Practicing in Interventional Pain and Rehabilitation Medicine

A common statement made by neurologists and ER, or trauma doctors is, "Every neck injury is a head injury."

However, less understood is, "Every head injury is a neck and spinal injury."

When people get their head struck, the neck and spine go along for the ride. The more accelerated the head movement, the more the neck and spine are affected. Concussions, in my opinion, are associated with neck and spinal injuries in most cases. What is less recognized is that many of what doctors thought were concussions (symptoms from the brain) are actually symptoms of the spine, referring back to the head much like an atypical migraine. I have named these symptoms, "Spinal Concussion Syndrome" (SCS). These indications are also seen in whiplash patients, or even in certain spinal patients with similar effects.

SCS is often caused by spinal cord irritation or impingement in the cervical or thoracic spine. It is associated with other symptoms as well. The term I created to describe the spinal injury and its symptoms are, "Spinal Myelopathic Syndrome" or "Tethered spine syndrome." The spinal cord is commonly being tethered by a disk or some spinal spondylolisthesis that is not recognized by most physicians, as of yet. MRI imaging grossly underestimates most spinal pathology, so it is simply missed, and can often only be found on detailed spinal examination. As a result of the under-diagnosing of standard MRI, I have created new MRI protocols for spinal imaging that I have titled, "The Whiplash MRI." This helps to identify more accurate details of injury in the spine. It may be also more helpful for improving the diagnosis of most spinal conditions.

Although SCS is a type of Spinal Myelopathic Syndrome, not all SMS has concussion symptoms, especially if the spinal cord impingement occurs in the lower part of the thoracic spine.

Common symptoms of spinal concussion syndrome are headache, jaw grinding and TMJ, sleep disruption, foggy head or fibro-fog, vertigo, tinnitus, fatigue, photophobia, visual auras, and even concentration difficulties. A concussion has very similar symptoms, but I will state many of the concussion symptoms that are assigned to concussion are NOT from the brain being concussed, but the spine becoming tethered. That is why many of the concussion symptoms can be delayed weeks or months. It is representative of the spinal injury advancing and not the brain. This does not take away from injuries to the brain from being concussed, but simply adds consideration to more advanced spinal rehabilitation to the rehabilitation of all concussion sufferers. In other words, to improve our concussion protocols, we must combine them with whiplash protocols, and vice-versa.

Anxiety is a common result in people with chronic headaches and spinal pain syndromes. Versions of post-traumatic stress disorder (PTSD) are frequently in whiplash patients. I can report improvement with advanced spinal rehabilitation. My explanation is that the neuropathic pain and spinal cord irritation aggravates or even creates some of the anxiety-provoking thought processes probably from a combination of factors, such as prolonged poor sleep, re-injury fear, constant pain, fear of retrogression, and a real concern that the medical profession fails to understand the patient's pain disorder.

Common causes of spinal concussion syndrome are car accidents, sports injuries, biking accidents, percussion bombs in war time. These events commonly involve sudden acceleration or deceleration, then abruptly stopping, causing the head to bobble on the neck and thoracic spine. It results in twisting and injuring the neck and spine. These whiplash effects damage deep spinal muscles and parts of spinal disks, causing progressive scarring and contractures in the muscles that gradually compress and twist the spine.

Often there is a delay of many of the symptoms of SCS due to the progressive nature of the spinal injuries. When the spine is whiplashed, it is rapidly twisted, compressed, flexed, extended, rotated in many directions. Disks are compressed, sheared, and torn. The intrinsic spinal muscles of the spine are injured and torn causing slow, gradual but nevertheless, progressive shortening and scarring of these muscles.

The effect of the scarred deep spinal intrinsic muscles is that they cause persistent spinal compression, rotation, subluxation, and scoliosis leading to progressive spinal disease. Common forms are disk compression, disk disease, disk herniation and sequestration, facet joint compression, spinal stenosis, disk herniation on spinal cord and spinal movement with vertebrae touching cord in under-diagnosed spondylolisthesis. The last two are more definitive causes of SCS when they occur in the cervical and upper thoracic spine.

Similarly, in simpler whiplash cases, the chronic neck and spinal pain often has delays of several months with progressive neck and back pain developing months or even years after the accident. This may be why some auto insurers try to get a whiplash sufferer to sign off on the accident early on, with cash incentives.

It is now well recognized that the pituitary gland is affected by concussions. The levels of hormones can be greatly reduced, affecting our total well-being. In cervical injury, the chronic craniofacial pain itself may also affect the pituitary gland's function and endocrine levels as well. Combined, these three factors, concussion, opiates and spinal injury have more broad effects than what we currently recognize.

The bigger picture is that we need to expand our spinal rehabilitation programs around the world. For best practices, these need to include:

- More detailed physical exam of the spine (currently quite poor or nonexistent)
- Detailed spinal imaging,
- Modified imaging protocols
- Improved MRI interpretation Expanded interventional spinal programs
- Combined spinal rehabilitation programs
- The addition of restorative technologies for the spine and brain

- Better sleep technology (current bed technology is 1960's) such that sleeping is truly restorative and therapeutic rather than aggravating; that would be a good start.

Over the past 25 years, I have created a number of proprietary, patented, advanced diagnostic, non-interventional and interventional spinal and limb therapeutic programs for neck, spine, craniofacial and limb pain and injury that I have employed for a number of pain disorders, including those with post-concussion symptoms. I will not discuss these specifically here.

There should be better recognition of the need for ongoing chronic concussion care, including ongoing rehabilitation. The current short-term passive care model is inadequate. The recognition that every head injury is a neck injury and spine injury is an important beginning. The neck or spinal injury may be the cause of many of the concussion symptoms. This fact needs to be better recognized and addressed for people to achieve the full potential of their journey through recovery to wellness.

G. BLAIR LAMB MD, FCFP
General Medicine
Practicing in Interventional Pain and Rehabilitation Medicine
Member of Ontario Pain Physician Section of OMA
Member of Canadian Academy of Pain Medicine
Patents in Spinal Rehabilitation

www.drlamb.com

BATTLEFIELD TBI

Andrew Marr
Warrior Angels Foundation
Co-founder and CEO

As a U.S. Army Special Forces demolition expert, I was in and around countless explosions. A career spent enduring concussive blast after concussive blast finally caught up with me. I began to suffer the effects of what we now know were multiple "minor" traumatic brain injuries. These were only minor in the sense that they did not put me in the hospital like any other kind of injury to the rest of my body would. It was hard to understand what had actually happened.

The "invisible" TBIs sustained in combat led to my medical retirement from the Army. These circumstances included the deeply affecting mental and physical manifestations of TBIs. It derailed my life, almost to the point of death. As a result of my injuries,

I became plagued by symptoms including:

- Depression
- Outbursts of anger
- Anxiety attacks
- Mood swings
- Memory loss
- Inability to concentrate
- Learning disabilities
- Sleep deprivation
- Loss of libido
- Loss of lean muscle mass
- Chronic pain
- Alcoholism

I developed a deep vein thrombosis (a blood clot that traveled and broke off into both lungs called a bilateral pulmonary embolism) and a number of other medically documented conditions. My drinking was so out of control that my wife, Becky, nine months pregnant with our fifth child, asked me if I could keep my drinking down to just one drink per day, in case she went into labor and could not drive herself to the hospital.

Things got worse until I confronted my inner warrior and explored my reasons for existence.

The "warrior few" are bound by a code. It includes a shared commitment and purpose found in teams whose code and mission are perfectly aligned. Their commitment to die, if necessary, is the ultimate form of love, strange as it may sound. It is a love for all we have trained for, bled for, and overcome along the way. Every last man's decision to commit fully produces immeasurable self-confidence and strength. The team must know that, if and when the time comes (and we all understood it may come for each of us); no self-reflection or contemplation is required. The commitment to act and sacrifice was made long before the soldier ever set foot on the battlefield. There is real liberation in upholding the warriors' code. It provides the freedom to live, serve, and love, without fear of failure, or death.

Warriors are united by an indelible bond. There is power in knowing that all move as you move, all think as you think, acting as a cohesive unit with a powerful, shared goal. Each compensates for the other in the places they may come up short. These unbreakable bonds are forged through our united experiences in combat. In Special Forces, we call it "The Brotherhood."

The battlefield is a place of unconditional trust and unquestionable sacrifice. It is its own world. It's where our thoughts sometimes go when we are physically in one place, but we know we really belong in another. It pulls us back.

Traumatic Brain Injury is called a signature wound of war. Had I not found the treatment protocols of physicians like Dr. Mark L Gordon, and Dr. Larry Komer, my future would have been bleak and I might have considered ending my life. This sadly, is what the wounds of war do, and have done, for so many service members and veterans.

I always subscribed to the principle that if you do not like something - fix it. When I reached out for help, the traditional response from the medical profession fixed nothing. My wounds were destroying my family and I. Our life together was unbearable.

The healthcare I was receiving merely was a series of mind-bending band-aids that did not treat the underlying condition. It masked symptoms and exacerbated problems. I told myself I was done with all of it. I was going out in search of a way to heal myself.

I decided to reject the life sentence of despair that the military medical model handed me.

The dedication to serve something bigger than myself repurposed me, saved my life, and illuminated a path that eventually led me to the answers in the comprehensive approach you find with Dr. Gordon and Dr. Komer in their clinics.

When this innovative treatment protocol was applied to me, I responded immediately.

Today, I am performing better than my pre-injury status. I have the privilege to speak all over the country. I enjoy running a successful organization. I have remained sober since October 2014, successfully quit tobacco, and have come off all 13 medications I was told I would need to be able to function.

My brother Adam Marr (former Army Apache Aviator) and I started the Warrior Angels Foundation (WAF) in January of 2015 to help other U.S. Service Members and Veterans receive the same level of care I received. I'm proud to say that WAF has helped over 150 U.S. Service members, and veterans receive the same life-changing protocol that so benefited me. They too, have rediscovered a quality of life.

This has allowed us to live a life of purpose and fulfillment, which can only be described as heaven on earth or the living embodiment of the motto, De Oppresso Liber.

WE MUST DISPEL THE MYTH OF "20/20 IS PERFECT VISION"

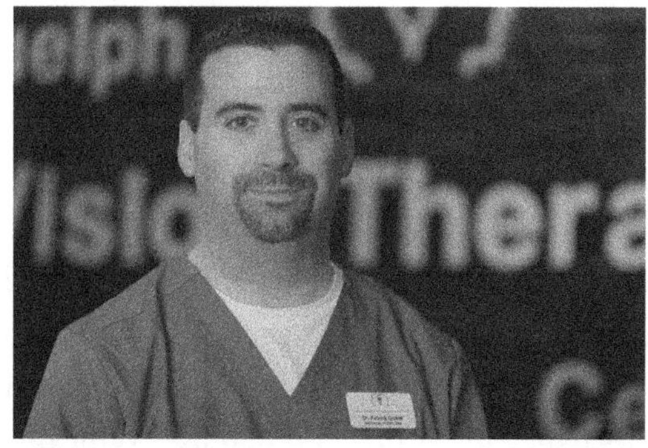

Dr. Patrick Quaid,
Neuro-Optometrist, FCOVD, PhD

Concussion is the classic "hidden injury." Many of my patients often state that they wish they had something "visibly wrong" with them so that family members and healthcare providers could realize that they are suffering, often desperately. Although the care of traumatic brain injury patients is slowly improving, many areas remain to be further investigated and researched. One such area is how the visual system is impacted in concussion. Its importance lies in the fact that at least 40% of the human brain is primarily visual machinery[14], which is astounding

14 "40% of the brain is primarily visual machinery"
Felleman & Van Essen, 1991 paper (abstract attached)

for just one sense (and is actually likely an under-estimation as only direct links were examined). Vision is essentially the "brain's way of touching the world" in that we use it not only to see detail, but we also use vision to determine where we are relative to other objects. In addition, we also use our peripheral vision to detect threats approaching, in addition to being able to react to such threats, as although the side vision is "blurry", it is very sensitive to motion in particular.

My own personal journey with respect to how vision is involved in concussion stems from an injury I sustained as a child at around the age of 8. I was hit by a car while playing on the road with friends (outside my aunt's house) and had a loss of consciousness for about 5 minutes. In Ireland at the time (around 1986), the overall management was (and arguably still is world-wide) nothing short of dismal. Standard operating procedure was to "observe" overnight. The advice really from there was about as clear as mud. More or less, if the patient is breathing the following day they are "fine." Of course, typically neuroimaging (even now) rarely shows issues and is not of much use. Typically, the issues that occur are termed "soft neurological signs."

Within about a week or two of the incident, I started to develop a speech impediment (moderate to severe stammering) and also some interesting visual issues. I would intermittently see horizontally double at near when reading (in fact I would just close one eye to make it easier to track) and my grades dropped substantially over the next 1-2 years. To be frank, the only reason I coped was the fact that my older brother (by 2 years) read to me. Thankfully, I finally came across someone who dealt with my speech issue and recognized there was also a visual issue and subsequently referred me to an optometrist who provided basic rehabilitative vision therapy (VT). The kicker? I saw three eye doctors who all said I was "fine", as I saw "20/20" prior. Funny thing is, we don't read at 20ft and all of my symptoms were at near. Nobody checked tracking ability or ability to convergence my eyes (pull my eyes inwards properly) and never mind I was light sensitive and that I closed one eye when reading! Far from normal from a visual standpoint, yet three times I was given a clean bill of health visually. Odd, right? Thankfully my issues were resolved within about 6-9 months with VT and I was back at honor-roll level at school.

Vision therapy has, thankfully, since then been absolutely validated (in 2008)[15] via a level one evidence trial (CITT study, NIH funded), which has absolutely disproved any suggestion that vision therapy is "hocus pocus." It is in fact, a highly effective method for resolving visual issues stemming from eye teaming problems. The brain controls the eyes. Given that at least 40% of the brain is primarily visual machinery, it is surprising that visual issues have been overlooked so much in the study of concussion. Studies in Brain Injury (world renowned concussion research journal) have shown clearly that not only is eye co-ordination affected in concussion but also that visual training can be highly effective in resolving many of the issues causes by concussions.[16] [17] [18]

Fast forward my life through an undergraduate degree, applying to an optometry school in the UK, and finishing the 4 year program with a first class honours degree. I was bitterly disappointed at the lack of training in the area of visual rehabilitation (termed binocular vision in University). Therefore, I subsequently left to study for my PhD in Canada (on illusions no less, essentially on how the brain is tricked by them and what this can reveal about the visual brain).

My PhD armed me with the knowledge to determine how the brain is involved in the act of seeing then likely how it is involved in concussion.

---

15 CITT study (subscript 2) attached

16 Oculomotor dysfunction and concussion association

17 Accommodative dysfunction in concussion

18 VT effective rehab in concussion

Better still, after training with COVD (College of Optometrists in Vision Development, a wonderful USA based organization that board certifies eye doctors in rehabilitative VT) I was now also educated on how to treat these issues. I was ready to help change people's lives.

Concussion patients who present to my clinic in Guelph typically present with the same constellation of symptoms visually. Again, this is ironic that there is actually little variance on the presentation yet few seem to connect the visual system to concussion. I will list them below for ease of reference (keep in mind I am not saying "some" patients, I am saying "most" patients):

i. Light sensitivity (photophobia)

ii. Motion intolerance (cannot tolerate peripheral motion, i.e. busy malls).

iii. Tracking issues when reading (saccadic dysfunction)

iv. Issues remembering what has been read (visual memory)

v. Intermittent double vision (usually more so at near or when reading)

vi. Difficulty reading / using a computer screen / watching a movie (large screen)

The interesting thing is that despite these issues, many (in fact most in my experience) of these patients still see "20/20" and the eye anatomically looks normal (i.e. no "disease" present.) As a clinician, it is vital to dig deeper to determine the issues causing the fairly consist presentation symptoms listed above. Tests such as the King-Devick, or the DEM (Developmental Eye Movement test) are very useful, and have been validated in concussion in determining if eye movements, termed "saccades", are abnormal. If such tests are not performed, quite frankly a patient being told they are "fine because they see 20/20" is not only clinically wrong, but also potentially devastating to the patient. Why "devastating" you may ask? If say a patient scores very high on a standardized symptom questionnaire (of which there are many, see our website for some examples, www.gvtc.ca, 0-60 scale, any score >15-20 a concern) and the doctor says "well, because you see 20/20 you are fine", the patient would be understandably confused. Essentially the only conclusion the patient is left with is that, "I must be nuts as no cause for my issues can be found." So, guess what happens the depression indexes of such patients increases (especially if family members start to "doubt the injury")?

If a patient has significant visual dysfunction and it is not addressed, look at the ramifications:

i. Issues using screens / computers / reading

ii. Issues coping with busy areas which may reduce ability to socialize in larger groups

iii. Difficulties in built up areas such as malls / grocery stores

iv. Tend to stay indoors and draw curtains due to light sensitivity

v. Tend to have issues with communication overall and issues can arise with close family members as they are perceived as being "fine." The Canadian Medical Journal has published statistics showing that in the concussion population there is a 2-3x higher rate of suicide compared to the general population. It is vital that healthcare practitioners (and the public) realize that these visual issues (in addition to all other areas discussed in this book) must be given proper consideration as quite frankly, if left untreated, can have devastating impacts on peoples quality of life.

Visual dysfunction is in fact highly treatable and we are proving it in our office in Guelph every day (as are many board certified Optometrists across North America and indeed around the world.)

We have also helped in research at the University of Toronto where we have baseline tested over 250 varsity athletes and found that 20-30% had issues on baseline testing (with respect to eye teaming skills). One of the key issues in the concussion world right now is, "how can two people sustain a similar injury playing the same sport, be the same age and gender, and yet both recover differently?" Of course, the answer lies in the pre-existing status of the patient. In our clinic, we frequently come across poorly recovering concussion patients who complain of academic/reading issues prior to the concussion.

Now, of course the concussion incident itself absolutely has a detrimental impact on the visual system, but it is not hard to see why a "weaker system" would be more susceptible to a more prolonged recovery. In fact, in conjunction with the University of Waterloo, we published information showing that children with poorer reading skills (on IEPs, or Individualized Educational Plan, which is essentially the schools terminology for "the kid needs extra help") have significantly poorer eye teaming skills. When these issues are treated, we see (in over 80% of cases) a dramatic increase in oculomotor function and "visual memory", which is critical for automatic spelling. When the cause of an eye teaming issue is not concussion, it can usually be traced to birth history factors such as prematurity (3 weeks or earlier), immaturity (low birth weight < 2,500g) or oxygen deprivation at birth (cord around neck, low APGAR). Therefore, children with these risk factors are not only more likely to have a weaker eye teaming system, they are very likely more susceptible to poor recovery if they sustain a concussion.[19] [20] [21]

19 Association between reading speed, cycloplegic refractive error, and oculomotor function in reading disabled children versus controls. Graefes Arch Clin Exp Ophthalmol (2013) 251:169–187

20 Variables associated with the incidence of infantile esotropia. Optometry (2007) 78, 534-541

21 An Exploratory Study of the Potential Effects of Vision Training on Concussion Incidence in Football, Optometry & Visual Performance, Volume 3, Issue

In closing, concussion primarily affects not the central vision (i.e. the high resolution area in our central 3-5 degrees of vision), but more the peripheral vision (or "side vision"). Dr. Eric Singman MD PhD (Neuro-Ophthalmology, Johns Hopkins) who is a valued friend and colleague, routinely uses a wonderful phrase for this issue, he calls is "visuo-spatial decoupling." What he means is that the central vision and the peripheral motion sensitive system essentially lose the ability to interact effortlessly. When this occurs, believe me, dizziness, nausea and poor overall mental ability (call it mental "fogginess) is inevitable. Rehabilitative Vision Therapy as provided by an optometrist with advanced training in the area (i.e. FCOVD Board Certification for example, see www.covd.org) is a vital factor in treating concussion. Of course, there is little doubt that "it takes a village" to treat concussion, but it is also quite clear that the visual system is a commonly affected area and needs to be addressed with a detailed work-up (and not just a "routine eye exam"). For more information on VT and how to find a doctor that can help, visit the College of Optometrists in Vision Development website (www.covd.org) and use the "Find A Doctor" tool (tick Board Certified box) to start the process. For more information and a free sign-up for a 2-hour lecture on concussion, learning difficulties and vision (given by myself at Hamilton General Hospital) go to www.gvtc.ca and click on the blue tab on the right side (just fill in the form and we will fire you off a free link).

There are also some questionnaires on our website that are free to use and they show a normal versus abnormal score (you can also submit the questionnaire and we can contact you with further advise if requested). The symptom questionnaires are all a 0-60 scale with any score over 20/60

2, April 2015 p 116-125. This paper essentially showed that rehabilitative vision therapy on one sports team and not the other results in a rate of 1.4 in 100 in the treated group compared to 9.2 in 100 in the untreated group ($p<0.001$). Good for thought! If we make the visual system better PRIOR to the hit, guess what, the player sees the hit coming faster and is able to GET OUT OF THE WAY or at least stand a better being a concern. There is also a research area and free PDF downloads (general information on VT) on the website.

If you or anyone you know has visual issues following a concussion (or a non-concussion, learning related reading issue) arranging for a rehabilitative vision therapy assessment for them may answer a lot of unresolved questions. Unfortunately, more and more patients/family have to become the advocates to ensure proper therapy (in the right sequence) is delivered to their loved one, the concussion patient. However, now more than ever, a team of professionals is required who actually communicate with each other. This is the real goal of this book and I applaud Dr. Komer for taking it on.

The first step however is public education.

Appointed Consultant Optometrist, University of Toronto
(David L. MacIntosh Sports Medicine Clinic)
Adjunct Faculty (Assistant Professor), University of Waterloo
School of Optometry & Vision Science
Chief Optometrist & CEO, Guelph Rehabilitative Vision Therapy Centre, Ontario, Canada

DOCTOR TO DOCTOR

This is a word to physicians and other healthcare professionals.

I have learned so much from patients who had concussions, TBI, and PTSD.

They are frustrated because they are not improving. They have "hit the wall."

Many have the impression that their health providers have given up on them. In fact, many of them expressed a fear that they would not get any better.

I know as physicians and professionals, we have to be realistic and truthful.

However, I also know that we have seen "miracles" over the years. One such story stands out for me and I want to share it with you.

I have delivered 13,000 babies. This was a huge part of my earlier career before I started concentrating on hormonal care. One night, when I was on call, there was an obstetrical emergency. The laboring mother started bleeding profusely and had a placental abruption. This is the situation where a portion of the placenta comes away from its attachment to the uterus and causes significant bleeding. In this case, the baby suddenly receives much less blood flow and much less oxygen. In fact, the baby may not survive the event. The mother is also at great risk. An emergency caesarean section was the only thing that could save the baby.

As usually happened, I had a great anesthesiologist and nursing crew working with me that day. Everyone sprang into action and moved the mother to the operating room and prepped her for surgery in record time. There was no opportunity to give an epidural or spinal anesthetic, so the mother had to receive a general anesthetic. There was no time to count the instruments. Every second matters during these critical situations. In less than two minutes, I made my initial incision and got the baby out within seconds.

I was crushed. The baby was floppy and deadly white. Throughout the hospital, a cardiac arrest call was made. We were blessed with having excellent pediatricians available for such emergencies. One of the very best, Dr. Bob Issenman, was already in the room to resuscitate the baby. He had his team of skilled nurses with him.

At this point, as the surgeon, things are very much out of my hands. I stood there and delivered the placenta and started sewing up the uterus. As the room was filling with people from the arrest call we had to tell them that we had all the help that we needed. As I looked over at Dr. Eisenman, who working furiously on this baby, hope was draining from me. The poor infant continued to be pale, floppy, and I did not hear any crying. The resuscitation continued for many minutes and still there was no sign of life in the baby.

I finished completing the C-section and said to Bob "It is time to let the baby go." This was the sign of defeat that no obstetrician likes to face. I was devastated.

However, Bob raised a finger. He looked at me and said, "I hear something." He was getting a faint heartbeat. Bob soon rushed the baby out of the operating room to the Intensive Care Nursery. The baby had survived the first half-hour of its life but was in critical condition. She was being transferred to the McMaster University Medical Center where

I had done my training. It was, and still is, known as being one of the top obstetrical and neonatal centers in Canada.

Later that day, when the mother was stable from of the operation, she was transferred to McMaster to be with her baby.

Over the next few weeks, I learned that the baby had survived and was progressing.

As time went on, I did not see the mother and did not have any follow-up information on baby M. Seldom a day went by that I did not have thought about this child. I was very aware that such critical situations can lead to cerebral palsy, brain damage and poor development. It haunted me.

It is interesting that obstetricians seldom think about all the babies they saved, but dwell on the ones that did not do well, in spite of the skill and caring of the whole obstetrical team.

Many years later, the mother was referred to see me for a gynecological problem. When I saw her name on my patient list that day, the whole situation of the delivery instantly came back to me. It was emotional.

We chatted a bit. I obviously felt somewhat reluctant to ask how baby M. had done. I wasn't even sure that the baby had survived these years.

I finally asked the inevitable question about the health of M., who would be nine or ten years old at this point.

Her mom's response left me reeling. "M. is wonderful! She has been identified as being gifted and has skipped two grades."

This did not fit into the science that I knew. It was against my training.

I was flooded with emotions. I was incredulous. I was happy. I was shocked. I was relieved.

Right after that visit I sent a note to Bob Issenman about the visit. Once again, I let him know my great gratitude and respect for the caring and skills that he displayed in saving this baby's health and life. Fairly soon after that I got a phone call from Bob, who was obviously choked up. He said that his tears stained the letter I had sent him. He also did not know the long-term outcome for this baby. He too was surprised and happy. We also discussed the fact that there was a message to both of us in this.

We learned that as smart as we think we are, there is much we do not know, and that we cannot predict the future.

What I also learned is that we have to take a very positive approach with our patients.

The patient wants to know how much we care before they care about how much we know.

We have to listen to their concerns. I tell my residents that we have to develop a good relationship with the patients. Part of that relationship is telling them we acknowledge their concerns and problems. I do not think that we can say things like, "All of your tests are fine. The problem is all in your mind." Most patients would not be in front of us, if everything were fine. This acknowledgment of the problem improves our communication and relationship with the patient. It also improves our level of care. Many times that acknowledgment breaks the ice and cements the relationship.

Patients are often frustrated because they have seen so many healthcare professionals, but still they are not improving. I always tell them that I am taking a totally fresh and different point of view of their problem. What I am doing is very unconventional compared with the traditional care for brain injury and PTSD. 56% of the patients after brain injury have abnormal pituitary function, and hormone levels at three months, and at 12 months, 36%. This means that a tremendous number of patients need hormonal screening and treatment.

In many of the patients I have seen with PTSD, there is a hormonal component, which is fixable. The science is very clear in our practice and will become more widespread with time. It offers new hope. I have found that hormonal restoration works in the majority of the patients who are referred to me.

The other thing that we can do if we cannot personally help the concussion, TBI, or PTSD, is to seek out those gifted individuals, such as the visual therapists, the audiologist and interventional pain specialists who can play a role in helping to improve recovery.

I also want to mention the assessment of depression. Each of the men we see in the Masters Men's Clinic completes a Beck Depression Index II. It startled me to see the depression rates in men sent to discuss testosterone replacement therapy. I looked at 41 consecutive new patients at our clinic and found that 37 had at least mild depression, with more than half having moderate to extreme depression. Only four had previously been diagnosed by their physicians. We have presented our work showing improvement of 68.4% in mood scores on the Beck, in the men that have treated with testosterone or thyroid replacement therapy.

What is the lesson from this? I think that we as physicians are often missing the diagnosis of depression in men. I have been surprised to see the scores showing significant depression in the men that I have just interviewed. From my training, it did not appear that they were depressed and yet the Beck shows that they are. And after treatment, Beck scores go down considerably. I think the men themselves are not really aware of their depression, and have other explanations for the way they are feeling. Perhaps they hide it well.

Another lesson is that many of these men would be better treated with hormones rather than antidepressants. Antidepressants have many negative side effects. In contrast, side effects of hormones are often positive such as decreasing heart attack, stroke, and more.

It has been said that medical knowledge doubles every 11 hours so do not sleep in, or you will be even farther behind!

It is difficult to keep up with all aspects of our work but with the World Wide Web, this has become much easier. I often tell my patients "I know a whole lot about very little." This specialization in one tiny area allows me to keep up in my subspecialty. If you find a passion in an area of your work, I urge you to go ahead and become more proficient in it. Medicine gives us many opportunities because the field is so wide. It allows us to renew our interest every day.

As a physician or health professional, you are incredibly well-trained and very bright. We all have stressful situations that affect us professionally and personally. This often detracts from our work satisfaction. Sometime, we have to step back, and take a wider view of what we do.

Enjoy your work and enjoy your patients. Cherish your colleagues who are skilled, dedicated, and who are great people. We have been gifted with the privilege to improve people's lives.

We will leave a legacy in lives touched.

Well done.

THE CHOICES ARE YOURS

I was uncertain what I would put in last chapter of this book. I knew it needed to emphasize the message of hope. I wanted to give you some clear action steps to take. Then yesterday, this chapter became crystal clear when Tony came to my office for a follow-up visit.

Tony is a 71-year-old man who I met for the first time six months ago. He was a referral to me by one of my classmates who is a family physician in a city a couple of hours away. Tony was in a serious motor vehicle accident in 2005.

Initially, Tony complained of fatigue, depression, and problems with concentration. I discussed with him that head injuries can lead to poor hormone levels. These hormone shortages could be responsible for some or all of his problems.

He had low testosterone levels. His total testosterone was 476 ng/dl (16.5 nmol/L) with the ideal over 808 ng/dl (28 nmol/L).

The rest of his hormone levels were fine.

Tony also had completed the Masters Andropause Screening Index©, a 50 question survey that we have designed to quantify the symptoms of low testosterone. He scored 180 out of 250, which was a very high score. He completed the Beck Depression Index II© and scored 22, which corresponds to moderate depression.

We gave Tony the choice of starting various kinds of testosterone replacement. He chose to go on injections. (As an aside, we have now trained more than 1000 men to inject themselves. Tony said that his son was a nurse who could help him, if there were any difficulties.)

Next, we started Tony on Vitamin D and Omega-3 fish oils to reduce brain inflammation and allow healing. Tony also had sleep disturbance. He agreed to take melatonin at bedtime.

I saw Tony only six weeks later, and he said he was already feeling better. He was happier. With better balance, he felt like going back to the gym to work out. Tony also found it easier to choose the right words to express himself.

At a later visit, we optimized his testosterone level by reducing his medication slightly.

Tony's most-recent visit yesterday was a revelation. He followed our treatment plan for six months. His improvement continued and he now felt that he was getting his old life back.

Tony was a stone carver prior to his injury. For some time, he was unable to return to his beloved craft. He then told me his story and showed me pictures. He carved an Orca whale sculpture. It was close to completion when the dorsal fin cracked and fell off. The broken piece ended up on his windowsill. He thought he might be able to save the sculpture by changing its shape to a porpoise. While carving the nose on the sculpture, that piece broke. Again, he put the broken piece on his windowsill.

Rather than getting frustrated, as he often did following the accident, Tony problem solved. He decided to carve a new dorsal fin from a totally different piece of rock and attached it to his sculpture. He fixed the nose then screwed on the dorsal fin. Once again; Tony had an Orca whale.

I was amazed. Tony had a great carving ability, but his accident prevented him from doing something he loved. The treatment allowed him to return to this work and reduced his frustration. That allowed him to persevere. Following the treatment restored his ability to problem-solve. His sense of humor was back. Now he could laugh at the problems he met. This allowed him to return to doing something he loved. Tony was able to carve and show me a tangible expression of his recovery.

It struck me that our treatment for concussions, TBI and PTSD was like carving a stone. Carving allows the artist to release the wonderful shape buried within the stone or rock. Similarly, our protocols and treatments cut away the layers of pain, disorientation, disability and depression that encase people following their injury. The beautiful person inside is once again freed to enjoy a fulfilling life, just as Tony has.

You may be one of these treasures buried within a post-concussion syndrome or PTSD. What will it take to see you free?

What choices can you make?

Forget about the doubters who said that you could not get better. Develop the positivity to allow you to succeed. Read other motivational books to build this positive attitude, such as Chicken Soup for the Soul.

Determine if previous investigations have included hormone levels. This might include having these results reinterpreted to see if they are optimal or not. Remember, the wide range that the lab says is "normal" is not truly the ideal or optimal range. Find a physician to help you interpret these and who will be able to prescribe treatment to decrease the inevitable inflammation that follows a brain injury. Do not forget, a brain injury may be a factor in PTSD. All of these conditions are improved by optimizing your hormones and supplements.

Set your goals and write them down. The best way to do this is to write a positive statement with a timeline. Be specific. Include a positive emotion in your statement. Instead of writing "I want to be happy," write, "I see myself on vacation in Sedona, Arizona happily walking trails by January of next year." Commit to your goals and visualize them each day. Success is more likely if you take this action step. Share the goals with other people in your life.

Make a plan. Surround yourself with people that you love and who love you. Write down the action steps that you need to take to be successful. Make a note of who in your support group can help you in each step. Write a timeline for when you plan to and proposed dates to complete each step.

If you know you have a problem with visual processing (such as difficulties with reading, light sensitivity, or eyestrain) find someone who is an expert in visual processing assessment. This may mean doing some research on the Internet to find help in your area. Perhaps your family physician or clinic can assist you.

Those with ringing in your ears (tinnitus), noise sensitivity, difficulty hearing or problems with your balance, see an audiologist. Search out one in your area.

If you have significant neck, back pain, or stiffness (especially neuropathic pain or burning pain), find an expert in interventional pain management. There are several kinds of treatments available, and each expert may approach treatment somewhat differently.

Of course, if you are getting other treatments (such as physiotherapy or psychotherapy or counseling) that are helping, continue them as needed. If you found no help with such treatments, skip this step.

We hope that this book gives you new insight into your challenges. At the clinic, we have cared for many people who are making excellent progress from brain injuries, concussion and PTSD. Now, you have new information and reason for new hope. You have an action plan, and you have help. The "old you" still is waiting inside ready to break out. We hope that your new path to wellness and vitality is successful. It is possible.

Choose your best future. You are worth it.

IMPORTANT INFLUENCES

Dr. Mark Gordon

Dr. Mark Gordon is a medical pioneer and brilliant scientist. He lives and practices in Los Angeles, California. He is a great physician, a gifted teacher and a loving family man. He invented the term Interventional Endocrinology to describe what he was doing. Over his 25 years of caring for patients, and much research, he developed an innovative way to practice his specialty of Endocrinology. He strives for optimal ranges of hormones, in contrast to the very wide ranges that were generally accepted, but that are far from ideal. He searched the scientific literature for natural products that would not only improve hormone levels, but would also improve lives. He integrated these into his practice of medicine.

He described the importance of hormones in many organ systems, particularly in the brain. His research allowed him to put forward the theory that not only was there damage from the initial brain injury, but there also was inflammation that could be more damaging down the road.

Other physicians were not practicing medicine the way Mark was. It is often difficult to change a colleague's minds to go in a different direction than the current standard care. When the usual treatments were not working, Mark took it upon himself to establish protocols that worked.

I followed a parallel course to Mark in my practice of medicine in Canada. I formulated similar ideal ranges of hormones and aimed for these levels in my patients. I became an expert in female hormones and wrote some of the Canadian guidelines and later did the same thing for men. I have been fortunate to practice Sports Medicine where I recognized abnormal hormone levels in my patients with head injuries.

Mark was ahead of me in the work he had done with traumatic brain injuries and hormones. I met him many years ago when I took one of his courses, and it was there that I realized that I too was an Interventional Endocrinologist. He is my colleague and teacher. We are friends. We share the same respect for patients and the joys of getting them better.

Mark is a world leader in this type of medicine.

He not only saw the hormonal changes in the brain injuries of troops coming back from combat, he made it his goal to improve their lives. Mark is a philanthropist. He crusades for the troops. His has attracted publicity regarding their difficulties. He has met many people in power, including former presidents and discussed the plight of military veterans.

Mark has made it his goal to train other physicians in Interventional Endocrinology and Neuroendocrinology (the study of brain hormones).

I applaud, congratulate, and thank him. His support and ideas have helped to make this book a reality.

DR MARK GORDON
The Millennium Health Centers Inc
16661 Ventura Blvd, Suite 716
Encino, California 91436
United States

www.millenniumhealthcenters.com

DR. JOHN CRISLER

John is a dear friend who I first met in a conference we were both attending. We started talking while waiting in line and it quickly became apparent that we shared similar ideas and value systems. John had specialized in men's hormonal health when very few others had. He was a leader in the United States.

We traded ideas about treatment and freely shared information that we had. I learned some new treatment protocols for John and quickly put them into practice with my patients. They worked very well.

Over the years, we have seen each other at meetings and each time it has been a joy to sit and talk. We not only discuss what we were doing in medicine, but also what we were doing in our personal lives.

John has improved the standard of care for men in the United States. He has a large following on his website for good reason. I know that I can phone John any time I have a question that needs an answer, and he will always give me excellent advice.

He has expanded his practice to include traumatic brain injuries. It is encouraging to see such a fine physician and person helping spread the word about treatment for TBIs. His great depth of knowledge can only lead to success and help many patients. I certainly can give Dr. John my highest recommendation.

DR JOHN CRISLER JOHN CRISLER DO
6425 South Pennsylvania Suite Five
Lansing, MI 48911
(517)485-4424

www.allthingsmale.com

Queen's University, Kingston, Ontario, Canada

I attended Queen's University as an undergraduate receiving a bachelor's degree in biology. I stayed at Queen's and was working on my Masters of Exercise Physiology degree when I was accepted into medical school there. My time during the four years at Queens School of Medicine was one of the best experiences of my life. I made many lifelong friends and received a great education. Their teaching allowed me to take the varied paths that I later followed.

I was also very fortunate to meet my wife, Joan, there.

I will always be a Queen's Man and will maintain a fondness for this great University.

Joseph Brant Hospital, Burlington, Ontario, Canada

I was on the consulting staff in Obstetrics and Gynecology for 39 years. I delivered 13,000 babies there. I carried out and an even larger number of surgeries. There are a very large number of physicians and hospital staff that helped and encouraged me throughout those years. My medical colleagues practiced superb medicine. The nursing staff always had my personal admiration and I owe them a great deal for any successes that I had. They were my friends and part of our team. I miss them since I have given up my hospital practice.

There are other members of the hospital team that sometimes do not get the appreciation that they should. I always tried to make them know how much I valued their work on "Team Komer." These include the administrative staff that checked in the patients and the porters who got them to where they needed to be. A special commendation goes to the cleaners, who often made a life-and-death difference when they quickly cleaned a C-section room between emergency deliveries so that the next baby in distress could be saved. As in any great team everyone had a key role, and all did a fantastic job.

McMaster University, Faculty of Medicine

After I finished my M.D. degree at Queens, I was fortunate to be accepted into the Department of Obstetrics and Gynecology resident program. I received superb training and experience for the four years in the program. McMaster was a relatively new medical school with modern ideas. I was the beneficiary of this excellent program. The physician professors there prepared me to practice my specialty but also built-in a lifelong thirst for knowledge and continuing education. They allowed me to think outside traditional guidelines. They taught me how to analyze the scientific information critically, and integrate the best of this information into my practice.

It was because of this training that I was prepared to develop many subspecialty interests that kept up my enthusiasm, during so many years of practice. The teaching was also an excellent basis from which to learn other areas of medicine. This often meant a journey down new roads that others had not explored.

I am grateful to the Department of Obstetrics and Gynecology at McMaster to have been appointed Assistant Clinical Professor in the Department and to continue having this role. I would like to thank Dr. Nick Leyland, Chief of the Department of Obstetrics and Gynecology, for continuously striving to improve the quality of care in our specialty and for his friendship and support over many years.

BOOKS TO READ

Rewire Your Brain: Think Your Way to a Better Life by John B Arden PhD (John Wiley & Sons Inc.)

The Brain's Way of Healing: Remarkable Discoveries and Recoveries from the Frontiers of a Neuroplasticity by Norman Doidge MD (Penguin Books)

Kids, Sports and Concussion: A Guide for Coaches and Parents by William Paul Meehan,III,MD (Praeger)

The Power of I Am by Joel Osteen (Faithworks Publishing)

Paths to Recovery by al-Anon Family Group (Al Anon Family Group Headquarters)

Drink by Ann Dowsett Johnson (Harper Collins Publishers Limited)

Opening Our Hearts - Transforming Our Losses by al-Anon Family Group (Al Anon Family Group Headquarters)

Courage to Change by al-Anon Family Group (Al Anon Family Group Headquarters)

Brave Enough by Cheryl Strayed (Alfred S. Knop)

Broken Open by Elizabeth Lesser (Billiard Books)

The Secrets of Maifesting by Wayne Dyer (Hay House)

Wishes Fulfilled by Wayne Dyer (Hay House)

The Success Principles by Jack Canfield (HarperCollins Publishers Ltd)

Thrive by Arianna Huffington (Penguin Random House)

Chicken Soup for the Soul: The Chicken Soup for the Soul Stories that Changed Your Lives by Jack Canfield (Simon and Schuster)

The Brain That Changes Itself: Stories of Personal Triumph from the Frontiers of Brain Science by Norman Doidge (Penguin Books)

Discovering Your Oasis – Hope for Weary Caregivers by William S. Cook, Jr., M.D. and Grant D. Fairley (Silverwoods Publishing)

Music at Hand: Instruments, Bodies, and Cognition by Jonathan De Souza (Oxford University Press)

Spark!: The revolutionary new science of exercise and the brain by Dr. John J. Ratey (Quercus Publishing)

Your Third Act - A Guide To A Great Retirement by William S. Cook, Jr., M.D. and Grant D. Fairley (Silverwoods Publishing)

CORPORATE WELLNESS & PRODUCTIVITY

Corporate wellness programs enhance the productivity and retention of employees. Senior management carries the extra burden of leadership stress along with the challenge of maintaining a healthy work/life balance.

Since 1991, the Peak Performance Institute Inc. has offered personal and professional development programs across the US and Canada. Our motto is, “Developing the Mind to Improve Performance.”

Our corporate performance programs include seminars, workshops, and retreats. The seminars share powerful information to enhance women’s health, men’s health and brain health. For individuals, a personalized wellness plan is a tool to maximize life now, and in the future. Executive coaching is also available to for personal and professional development.

For governments, we offer to consult on policy and programs for concussions, TBI, and PTSD. We are also a resource for men’s health and women’s health.

Corporations, governments, educational institutions, and organizations are invited to contact us to discover how we can maximize the performance and well-being of your executives and entire corporate team.

For more information, email us at info@peak-performance-institute.com

HELP AND HOPE FOR VETERANS

The men and women serving the military to protect our freedom deserve our respect and gratitude. They are also entitled to the very best in health care when they have suffered an injury.

As you have read throughout this book, some injuries, like concussions, are cumulative. Some are also difficult to identify using many conventional tests. Left untreated, concussions, TBI, and PTSD can lead to a life of despair and isolation.

Many veterans believe that they are beyond hope. The life they have now is as good as it will ever be. However, there is new hope for our servicemen and servicewomen. With the assessment, diagnosis, and treatment solutions now available, a better present, and future are possible for many of our military personnel that suffer these injuries.

The Komer Brain Science Institute is ready to consult with government, and other institutions, to promote effective treatment of our veterans. Our consultations also provide new insights on concussions, TBI and PTSD to aid the military with an enhanced understanding of minimizing, assessing, and treating personnel while they are serving. Early intervention can greatly minimize the long-term consequences of brain injury.

Contact the Komer Brain Science Institute for more information.

Email info@komer-brain-science-institute.com

SPORTS CONSULTANT

Dr. Larry Komer has a long history of involvement in sports. With his research and application of hormone restoration and supplement optimization, he is in a unique position to consult with individuals, organizations, and corporations and to make a significant improvement in athletes.

He was a hockey goalie, playing on high school, university and adult teams for 25 years, following his graduation. He contributed to innovations in goalie equipment during his playing career. Besides playing, he served as a trainer and physician for many teams. He also played lacrosse, and for decades has been involved in the sport. He has served as a coach, trainer, and team physician to many teams in minor lacrosse.

Dr. Komer was the team physician for the Brampton Excelsiors in Major Series Lacrosse in 2016. Dr. Komer continued on the Medical Staff of the Toronto Rock Professional Lacrosse team for more than 15 years. The team has won six titles in the National Lacrosse League. He was Team Physician for the Team Canada Lacrosse Team when they won the silver medal in the Women's World Outdoor Championship in 2013. Dr. Komer was Team Physician for the Team Canada Men's Lacrosse Team when they won gold medals in 2003 and 2015 in the World Indoor Lacrosse Championships. As well, he has been Team Physician for the Hamilton Nationals in the Major Lacrosse League.

In his private practice, he has treated both male and female athletes, in a variety of professional and amateur sports. He is seeing an increasing number of professional athletes in contact sports such a football, hockey and lacrosse and is helping them return to the game they love. Dr. Komer is a consultant to the International Lacrosse Federation (FIL).

He has a keen interest in healing athletic injuries as well as injury prevention. He has a great knowledge of supplements. He has applied this knowledge of hormones by inaugurating the Masters Men's Clinic in 2004, where more than 6000 men have been treated. He founded the Komer Brain Science Institute in 2015 specifically to treat concussions in both men and women and to try to reduce long-term illness associated with brain injuries.

Dr. Komer has lectured nationally and internationally to a large number of organizations. He has worked closely with professionals in a variety of disciplines and consults with a team of innovative and passionate individuals who care keenly about the care of athletes.

Dr. Komer's latest project is assessing and instituting a supplement program to optimize brain function and performance in high performance athletes. These supplements will also lessen the effect of future concussions as well as help them heal more quickly.

For more information on how Dr. Komer can assist you or your organization, contact him at doc@drkomer.com.

CONSULTING SERVICES

The Komer Clinics consult with corporations, insurance companies and the pharmaceutical industry to develop wellness programs and to further research into the applications of The Komer Method to various areas of healthcare.

As a physician with 40 years of experience within the field of Obstetrics and Gynecology, with particular expertise in hormonal therapy in both men and women, and now with treatment of concussions and traumatic brain injuries, Dr. Komer is in a unique position to assist corporations and organizations in their work.

Dr. Komer combines clinical experience in both the hospital and office context, as well as his work as a seminar instructor to provide insights on both men and women's health, in particular, and healthcare in general.

Throughout his career Dr. Komer has continued his professional development through conferences and educational events.

Dr. Komer has provided leadership and served in a number of professional associations and boards in the healthcare field locally, provincially, and nationally.

Writing for both popular and professional publications reflects Dr. Komer's commitment to assisting both the public, and other health care professionals in their understanding of women's and men's health. Dr. Komer is skilled in preparing educational events, being a lecturer and a facilitator in many of these.

Dr. Komer has had extensive media training. He has been in on the product launch of several new medications, including very successful press conferences and media days. Three of these medications were Alesse, Estrogel and Prometrium.

One of Dr. Komer's fortes is preparing a pharmaceutical sales staff for interaction with physicians about their pharmaceutical products. He takes a "real-life" in contrast to a theoretical approach, so that pharmacy reps are well prepared to detail their product.

THE KOMER WOMEN'S HEALTH CENTRE

The Komer Women's Health Center began in Burlington, Ontario, Canada in 1976. Dr. Komer was on staff at the Joseph Brant Memorial Hospital in Burlington for 39 years as a specialist in Obstetrics and Gynecology. During that time, he delivered more than 13,000 babies and carried out even more surgeries.

Initially, Dr. Komer cared for a large number of high-risk pregnancies since he was the most recently trained and youngest member of the department. Then through training with Dr. Kurt Semm, he learned advanced laparoscopy, which allowed the hospital to do surgery through a small telescope and save the patient major surgery. Like many who invent new methods or understanding in healthcare, Dr. Semm's fresh ideas were not widely accepted, and he was ridiculed for years. However, in 2002, Semm received the "Pioneer in Endoscopy" award in recognition of his contribution towards the advancement of many surgical specialties. His insight and persistence changed medicine for the better, improving outcomes for patients.

Dr. Komer developed an interest in the treatment of infertility. He and Joan felt fortunate to have two children, and he wanted to help others have the opportunity to experience the joy of having kids. This required an extensive knowledge of hormones and was the foundation of much of what he did in the rest of his career. Over the years, he had the honor of having four girls called (Lauren or Laura) named after him. With such a long practice, many of the women who he cared for through infertility and in pregnancy still see him in his gynecologic practice. They are like family. Dr. Komer enjoys when many of these patients show him pictures of their children growing up.

He became one of the first in Canada to train in operative hysteroscopy. This is procedure using a small telescope to enter the uterus to perform surgery and avoid a major operation. Once again, he is grateful to be trained by a brilliant gynecologist, Dr. George Vilos, who continues to be a leader in Obstetrics and Gynecology working at Western University in London Ontario. He is a great colleague and friend. Dr. Komer carried out more than 7500 of these surgeries in his career and trained many residents this technique.

As time went on, many of the women he cared for during pregnancy became menopausal. He spent time researching menopause and hormones. Quickly, he understood the horrible symptoms from lack of hormones as well as the increased risk of serious medical problems. Long before it was popular and accepted, Dr. Komer became an advocate for women at this stage of their story. He helped launch the first commercially available pharmaceutical grade bio-identical hormones and lectured extensively about their benefits of bio-identicals over the hormones that were commonly used at that time.

It is amazing that he still hears arguments about restoring the hormones that have been a normal part of every woman's system for almost 40 years of her life. It makes sense that if the high levels of normal hormones women have are safe beginning at the time she starts periods until menopause, then small levels of the same hormones should not only be safe but helpful too.

Did you know that no animals reach menopause? The majority of women did not live to reach menopause until the 1900s. So the argument you hear that menopausal women have lived without hormones for thousands of years is erroneous. In fact, humans were not designed to outlive organs such as the ovaries. However, with safer water, better nutrition, medications such as antibiotics, and medical solutions such as surgery, people are living much longer. There are receptors in almost every cell in the body that require hormones to run optimally. Dr. Komer's work has been to restore safe hormone levels in women who have run out of them naturally. Scientific articles show that hormone replacement reduces the risk of stroke, heart attack, diabetes, Alzheimer's disease, dementia, bowel cancer and osteoporosis. Along the way, Dr. Komer took advanced training in osteoporosis and befriended two of the Canadian leaders on this subject, Dr. Rick Adachi and Dr. Bill Bensen.

Some people believe that hormones increase breast cancer rates. A study by Fournier of 80,000 French women for 8.1 years showed a 10% reduction in breast cancer rates using bio-identical hormones.*

There are 17 symptoms of menopause that hormone-replacement therapy reduces or eliminates.

These are:

1. Hot flashes
2. Night sweats
3. Poor memory
4. Poor mood
5. Poor concentration
6. Poor energy or fatigue
7. Vaginal dryness

8. Bladder issues
9. Irritability
10. Depression (which can triple in menopause)
11. Joint pain and muscle ache throughout the whole body. (This is the only symptom that 100% of menopausal women get.)
12. Sleep disturbance
13. Decreased sex drive
14. Palpitations of the heart
15. Headache
16. Weight gain
17. Brain fog

Dr. Komer has seen more than 13,000 women for menopausal therapy. He wrote guidelines for the College of Family Physicians of Canada in 2009 regarding hormone-replacement therapy. He has lectured on three continents regarding menopausal therapy in women.

Most of the women he looked after felt incredibly better and healthier as they continued hormone-replacement therapy.

For more information on the Komer Women's Health Center visit

www.komer-womens-health.com

@womenshealthdoc

THE MASTERS MEN'S CLINIC

The Masters Men's Clinic is the largest clinic in Canada, diagnosing, treating, and doing research on testosterone deficiency syndrome (Low testosterone). The clinic has now assessed and treated more than 5000 patients, restoring vitality, happiness, and health.

Most of the women Dr. Komer treated felt incredibly better and healthier when they continued hormone-replacement therapy. Some asked if there was something he could do for their husbands who seemed to suffer from many of the same symptoms. Their persistence with more and more asking for help for the men in their lives piqued his ever-present curiosity. It quickly became apparent that men could also suffer from the condition of low hormones. Like so many areas of healthcare, it is easy to continue in the busy stream of your practice and not recognize the cross-over of your knowledge to different challenge. Why would only women be affected by low hormones and not men?

So began his quest to understand about hormones in men. This led to a journey investigating Andropause; something has also known as Male Menopause, Low T, or Low Testosterone.

His curiosity and openness to learn were rewarded.

Dr. Komer quickly began to meet the people that would help him become an expert in Andropause. These important influencers included Dr. Malcolm Carruthers of London, England and Dr. Al Morales of

Kingston, Ontario, who are both pioneers in the field. They have helped shape his knowledge and understanding to where he realized he was not just a gynaecologist, but rather someone who understood the important role hormones plays in our lives.

He recognized that testosterone in men was very similar to estrogen in women. He used the concepts that shaped his knowledge of female hormones for 25 years and applied them to men's hormones. It worked! It helped him avoid the many blind alleys that would not be beneficial in treatment and instead led him to solutions that were just as successful in men as they were in women.

Dr. Komer's good friend and colleague, Dr. Mike Greenspan, a noted Urologist at McMaster University, Hamilton, Ontario supported his treatments and offered encouragement in his journey. Dr. Komer is eternally grateful to him for his continued support and guidance.

Dr. Komer has now seen and treated more than 6000 men in the Masters men's clinic in Burlington Ontario where he works with his friend and colleague, Gordon Tonnelly. Gord counsels and encourages men and consistently provides Dr. Komer with the latest research in the field of men's health. Gord was an integral part of getting the clinic started and for much of its success. Dr. Komer and Gord make a great team.

For more information on the Masters Men's Clinic, visit our website.

www.mastersmensclinic.com

@MastersMensDr

THE KOMER BRAIN SCIENCE INSTITUTE

Normal hormone levels are mandatory for optimal brain function. Disease, injury, stress and aging can lead to poor hormonal function. With a long history of studying and treating hormonal deficiencies in both men and women, Dr. Komer has witnessed the improvements that hormone restoration leads to in improving brain function. His special area of interest is concussions.

Dr. Komer has been restoring hormones in brain-injured patients for many years. He has a long affiliation with athletics. His original training was in exercise physiology before he became a physician. He has looked after athletes his whole medical career. Dr. Komer has been Team Physician for the Toronto Rock Professional Lacrosse Team since its inception 17 years ago. They have won six league championships. He has been Team Physician for Team Canada Lacrosse in The World Lacrosse Championships (2003 and 2015 when the men's team won gold medals and 2013 when the women's team won their silver medal). Dr. Komer has been Team Physician for many other lacrosse teams both in Canada and in Europe. With these teams, he has seen and treated many concussions.

He is a colleague of Dr. Mark Gordon from Los Angeles who was the first in the world to present a formal three-day seminar on traumatic brain injuries and the association with disturbed hormone levels. Dr. Komer has led a parallel life with Dr. Gordon. Dr. Komer's expertise was developed with athletes whereas Dr. Gordon's was developed looking after soldiers in the US military returning from battle. They have discovered very similar findings and developed treatments. Dr. Komer was pleased to be accredited by successfully passing this course and its exam. He holds Dr. Gordon as a dear friend and a brilliant colleague. They both agree that there has been a need to fill this treatment gap not only in service men and women, first responders and athletes but in the rest of the public as well. They are seeing more and more young people getting concussions and the scientific information now shows that even if the initial damage is mild, there can be inflammation, which may spread over the next two or three decades and lead to brain problems.

Dr. Komer's present protocols address and treat the inflammation as well as restore the hormones. Most individuals and clinics treating concussions and brain injuries do not do this.

The Komer Brain Science Institute does not feel that any one person is capable of treating all the ramifications of concussions and brain injuries. We have therefore formed a group of passionate and brilliant individuals with other medical expertise in a SuperClinic that includes treatment for:

- Vision and brain visual-processing problems
- Hearing and balance abnormalities
- Pain and muscle problems particularly in the neck and back
- Sports Medicine

Utilizing these professionals, Dr. Komer has found that our success rates have been much higher and patient improvements have been far superior.

For more information, visit our website.

www.komer-brain-science-institute.com

@KTBIDrkomer

BE MENOPOSITIVE! SEMINAR

Dr. Larry and Joan Chandler Komer have presented this dynamic and transformational seminar over 100 times in nine provinces with over 35,000 attendees. They have donated the more than $35,000 that was collected as the voluntary admission fee to women's shelters in the cities where they have spoken. The attendees rated the seminar "very good" or "excellent" in the 98% of the satisfaction questionnaires.

Joan Komer trained with Mark Victor Hansen and Jack Canfield, of "Chicken Soup for the Soul" fame. Her entertaining and engaging style is motivational and encouraging. Being in menopause herself, she is an excellent role model for other women in the same phase of life. Her message is about positivity and hope. Her practical examples and suggestions on how to thrive during the menopausal years apply to every stage of life.

Dr. Komer's portion of the seminar presents the very latest and accurate scientific information concerning menopause in a form that allows those in the audience to understand the topic, to discard the many myths concerning menopause and to take charge of their personal healthcare. Dr Komer discusses why hormones that are identical to those that women made for all the years before menopause are the ones he uses to ensure health and vitality. He translates what he has learned in 40 years of practice and outlines a template to guide women safely through this very important and rewarding time of their lives.

BE ANDROPOSITIVE! SEMINAR

This is the seminar that Dr. Komer has presented to many public groups as well as physicians and other healthcare professionals. It takes a very positive look at Andropause or low testosterone levels. The seminar addresses the major benefits of testosterone replacement in men who have low testosterone and are feeling fatigued, sleepy, sore, and lacking the vitality that they previously had. There is much controversial information about low testosterone, but this seminar cuts through all the inaccurate media presentations as well as many incorrect concepts generally held.

Dr. Komer points out the serious health issues, such as an increased risk of stroke, heart attack and depression that go along with low testosterone. He talks about the particular need of testosterone in the brain and the heart. Dr. Komer presents his own results on treating depression with testosterone in the men he has seen. There is an average of 68.4% improvement of their moods without antidepressants. Heart attack and stroke are substantially reduced and a summary of 40 years of scientific research are discussed to prove the point. He presents the very latest information so that it is easily understood, and it allows each man to customize his path back to better health and happiness.

BRAIN HEALTH SEMINAR

A Paradigm Shift in Treatment for Traumatic Brain Injury: Hormonal Restoration

Many patients with traumatic brain injury (TBI) hit the wall in their recovery and do not get completely better. Within three months of a concussion, 56% of patients have abnormal hormones and this remains at 36% at one year. With more than one concussion, the rate of abnormalities is much higher. However, traditional therapy does not even consider hormones let alone treat them.

There is a large amount of scientific information regarding hormones in brain injuries that has been scattered throughout the literature over the years. Dr. Komer is one of the few physicians who has applied this new science, using his expertise in hormonal restoration, to help brains heal. Looking back on the men that he is treated in the Masters Men's Clinic over the last 14 years, he has found that more than 80% have sustained a concussion, or brain injury prior to being low in testosterone and other hormones. The same process occurs in women with TBI, and the hormonal disruption does respond to hormone replacement.

Dr. Komer has been heavily involved in sports medicine for his whole career. He has been Team Physician for: the Toronto Rock Professional Lacrosse Team for 17 years, World Championship teams for Team Canada Lacrosse for both men and women and many other lacrosse teams. He has applied this knowledge of physiology and treatment to help restore their optimal brain function. There is a tremendous need that this information be spread so that more patients can be helped.

He has presented this seminar to both medical professionals and the public and received much feedback on how this new information has brought hope to many who felt that there was nothing further to help them. The highlights of this seminar were recently published in the Ontario Brain Injury Association Review magazine.

THE KOMER METHOD

The CORNERSTONE of The Komer Method is the attitude that only OPTIMAL is acceptable.

This encompasses optimal levels for hormones, blood chemistry, nutrients, supplements, nutrition, exercise and behavior in each individual.

Many lab ranges are very wide and are accepted as "normal." Large portions of these ranges include levels that are far from ideal. The Komer Method has developed its own set of optimal ranges, and strives to achieve these for each patient.

When physicians treat abnormal blood sugars, high cholesterol or high blood pressure, they choose the ideal standard as the levels for a healthy young adult. However, when the same physicians correct irregular hormone levels, they do not follow this practice. In fact, they will accept aging and deterioration of hormone levels as a normal event. Dr. Komer's belief (and what has worked so successfully in his practice) is that achieving these optimum levels at any age fine tunes the body to minimize the effects of age, time and stress. These levels result in men and women who are their healthiest. They lead to a reduction in long term illness and an increase in well being.

These optimal values have been developed over years of experience, through research by Dr. Komer and others, and add also patients reporting back what makes them feel their very best.

The Komer Method has developed its own protocols for diagnosis and treatment for various conditions, and these have been tested and improved in thousands of individuals. Innovation and continuous improvement of all protocols is ongoing. Medical literature is reviewed daily to integrate new research into The Komer Method.

A major emphasis of the Komer Method has been dedicated to achieving ideal hormone levels in both genders, for such conditions as menopause in women, and low testosterone in men.

Dr. Komer also treats other individuals, including both pro and amateur athletes, suffering from concussions or injuries. He has a particularly large percentage of men and women, who are in the military, are police officers, correctional officers, or firemen. These professions involve stress and long hours and sometimes trauma, which can lead to abnormal hormone levels. There is a need for understanding, assessment and treatment of these individuals.

Dr. Komer has been an innovator in bringing new techniques and fresh ideas to medicine and has, at times, stood alone in championing ideas that have turned out to be leading edge concepts in medicine. In his practice, he has over 13,000 women and 6,000 men in his program who are reaping the benefits of hormone restoration. They are feeling well and happy, functioning optimally and staying healthy.

ABOUT THE AUTHORS

Dr. Larry Komer and Joan Chandler Komer

Dr. Larry Komer is a sought-after expert and authority in restoring hormones, health, performance, happiness and vitality. This new area of medicine is called Interventional Endocrinology.

He has pioneered several treatment methods recognized as leading-edge protocols for the treatment of menopause, Andropause (low testosterone) and traumatic brain injury.

Larry Komer trained at Queen's University, Kingston, Ontario with a degree in physiology (the science of how living organisms function). He went on to receive his MD degree at Queen's and then went to McMaster

University, Hamilton Ontario where he earned his specialty degree in Obstetrics and Gynecology. Dr. Komer is Associate Clinical Professor at the Michael G. DeGroote School of Medicine, McMaster University, Hamilton, Ontario.

He joined the staff of the Joseph Brant Memorial Hospital in Burlington, Ontario in 1976 and since then has delivered almost 13,000 babies. He introduced many new gynecologic techniques to the hospital including operative endoscopic surgery, operative tubal surgery and reversal of sterilization, gynecologic laser surgery, diagnostic and operative hysteroscopy and founded the Bone Health Clinic during his research in osteoporosis. His office practice has included a wide range of gynecologic assessments, but in the last 20 years has focused on hormonal evaluation and hormone-replacement therapy (Interventional Endocrinology).

Dr. Komer has been a leader in the treatment of menopause, writing guidelines for physicians and delivering more than 100 talks entitled "Be Menopositive" with his wife Joan. He has assessed and successfully treated over 13,000 women who are suffering from poor hormonal function.

In 2004, he became Founder and Director of the Masters Men's Clinic. This is the largest clinic in Canada diagnosing, treating, and doing research on testosterone deficiency syndrome. The clinic has assessed and treated more than 6000 patients. Dr. Komer has been on the Board of Directors of the Canadian Society for the Study of the Aging Male.

During that time, he observed the link between traumatic brain injury and low hormones in both men and women and has been a leading advocate for hormonal restoration to re-establish normal brain function. In 2016, Dr. Komer was the recipient of the prestigious Physician Care Award.

Here are some of his reflections on his journey of discovery.

Some days I asked myself: "How did I end up here, doing what I do?" It is also a question that many other people have asked me. At times, it has seemed to be a strange journey.

I grew up as a typical Canadian kid in a happy middle-class home. I enjoy sports, and I enjoyed school. In school, I was usually the class comedian but seldom got into trouble for it because I did well academically. In high school I realized that I wanted to become a physician. There was no single event or person who inspired me to do this. My parents could not afford University when they were young, but they emphasized to my brother, Wayne, and me that they would make sure not only could they fund university, but also that education was a priority and we were expected to continue.

I played a lot of sports growing up, but hockey and lacrosse were my favorites. I went so far as to set up a hockey league in our high school because there was not one. Many of us played on teams outside of school.

My father worked at Thompson Products. The St. Catharines TeePees (named after sponsor Thompson Products) were the junior farm club of the Chicago Blackhawks. Growing up I met many junior hockey stars prior to the time that they became regulars in the National Hockey League. This included Phil Esposito, Bobby Hall, Stan Mikita, Roger Crozier and Denny DeJordy. The last two are favorites of mine. They both spent time with me and gave me some of their old sticks which were much too big for me but which I cherished deeply. They both became goaltenders in the National Hockey League. For some perverse reason, I wanted to be a goaltender, and I played that position from my youth in St. Catharines, through my time at Queen's University, during the time that I was training as a physician and a specialist right up until I retired from Old timer's hockey when I was in my 50s. It brought me great joy for many years, but I retired when the pain became worse than the pleasure.

I also played lacrosse, and it became my favorite sport. I coached or was the team trainer and later the team physician for many different teams. I personally suffered the usual injuries that occur along the way in these two sports.

After high school, I enrolled in Queen's University, Kingston, Ontario and earned an undergraduate degree in physiology. While I was working on my Masters in Exercise Physiology, I was accepted in Medicine and stayed at Queen's and received my MD. I then went to McMaster University School of Medicine to become a specialist in Obstetrics and Gynecology. This career choice was inspired by a kind and brilliant Obstetrician and Gynecologist in my training, Dr. Tony Daicar to whom I am very indebted. He helped me set up a rotation in Bradford, England during my training to learn from the Brits. Tony met his wife Maggie there, and they were married for 51 years. I had never told him that he shaped my career choice, and when he retired, at the urging of my wife Joan, I returned to Kingston and told him how he inspired me. At that point, I was the head of all the Obstetricians in the province of Ontario. He beamed and introduced me to all of his family. I was so glad I took the day off work to make the trip. We had a few encounters after that, and he treated me like a son. He passed away in April 2016 after a distinguished career and a very full life.

Dr. Komer's CV is available on his website www.drkomer.com

JOAN CHANDLER KOMER

Joan Chandler Komer is the CEO of the Peak Performance Institute Inc. She is an educator, self-esteem coach, seminar speaker, author, and a researcher. With Dr. Larry Komer, they have shared their "Be Menopositive!" seminar on women's health and well-being to over 35,000 people.

As Joan reflects, "I was youngest of three and a baby boomer, with two boys before me. I was the girl my parents so badly wanted. I was "a youngest child," but because I was 6 and 7 years younger than my brothers, I have characteristics of an only child too.

I always loved crafts. When I was in high school, I was in the "academic" stream, so I was only allowed to take Home Economics for the first year. Devastated, I arranged with the sewing teacher to audit all the daily sewing classes for three years. After school, she would show me what had been taught that day, and I would go home, do it and bring back the finished project, crafts and sewing, the very next day. Eventually, I made all my own clothes and sewed for my mother as well. I planned to take a degree in Home Economics, but changed my mind in Grade 13, as I wanted my passion to be a hobby and not a vocation.

I loved high school. I was a cheer leader, in the choir and played second saxophone in the band. I was in musicals. I won the Grade 13 music award for the highest marks in music. I was winning my campaign for President of Student Council in the 12th grade when I became very sick with mononucleosis and ended up spending the next 4 months in bed. Against my doctor's wishes, I went back to school for the fall for my Grade 13 year. There was no way was I going to miss moving on with my friends. I attended about half of the classes because I was so tired, but I did graduate as an Ontario Scholar.

Doing well academically was expected. I did, but I never felt as if I did, because I was always comparing myself to my older brother, David, who was an academic phenome. As a result, I developed a fear of speaking up in class, lest they find out the truth that I knew that I really was a fraud, not smart at all. It developed into a crippling fear of speaking in public that haunted me for a lot of my adult life, especially in University.

I received an honours degree in Psychology from Queen's University in Kingston. I won the Psychology prize in my third year, which surprised everybody, because I was so quiet in class. Instead of carrying on with Master's degree in Psychology, I enrolled in McArthur College of Education at Queen's, and with a degree in Education, Counseling and English, I taught high school for several years, counseling being my passion.

Music has always been very important to me. I have sung in my church choir for 25 years. I have performed in musicals and concerts. I love to sing, and I love to perform. I discovered later in my life, that I also love to act. This has led me to have leading roles in many amateur productions. Acting is energizing for me. It renews my soul, and has helped me cope through some of the very difficult times we have had in the last 10 years. I also enjoy speaking, as it is a way for me to perform.

I love flowers and color and beautiful things. Gardening has been a passion all my life beginning in my early years. Back then, I would spend hours with my father in our half-acre property. We planted every vegetable, fruit and flower possible. While my brothers liked to help our mother with the inside chores, I was the happiest outside. As a five-year-old, I spent all day outside, and my most memorable times were riding on the back of our tobacco-growing neighbor's cultivating machine. For years, I lovingly tended my 100 varieties of roses at our present home. Eventually, time and energy and the rose diseases made me change to less demanding perennials.

My early interest in crafts and sewing developed into an interest in and aptitude for interior design. I was one of those original Pinterest people. After 10 years of collecting ideas in a scrapbook, Larry and I designed and built our unique "California" inspired house, 35 years ago. I was inspired to put a garden inside down the center of the house.

More important than anything else I have done, I was a mother of two. I loved being a mother very much and gave up teaching to stay at home with our daughter Kim, and later our son Scott. I made the costumes for the many musicals and skating shows for both kids. Halloween required my creative sewing as they were very particular about the authenticity of their costumes. It was before ready-made costumes of today, and even the large craft stores. I attended every one of Kim's dance and musical performances, and hockey or lacrosse game of Scott's. Larry was a very busy Gynecologist during those days, so my committed role as a full time wife and mother was crucial, and never regretted.

Once the kids were in school and at an age to need me less, I found I needed more. I missed the high school kids, so for several years. I became a leader of a teenage girls' group at our church called Canadian Girls in Training.

As part of my searching for self-fulfillment, I had the opportunity to attend a conference called The Million Dollar Forum. It was an uplifting week of powerful motivational speakers. This included Jack Canfield and Mark Victor Hansen, authors of "Chicken Soup for the Soul®" series. As a result of my experience there, I arranged to attend training on Self-Esteem by Jack Canfield in California. It changed my life. I returned home to speak on Self- Esteem to groups. I also developed a seminar for women called "The Rose" (Radiant – Opulent – Self-Esteem). It was extremely well received, as low self-image is rampant in women, especially.

As doctors were not allowed to incorporate, I became the CEO of a management company, Anigav Professional Services Inc. that handled the business end of Larry's practice. (This was during the very early days of computers). We started Peak Performance Institute. We organized weekend seminars in exotic destinations for the personal development of physicians, and their spouses called "Chicken Soup for the Physician's Soul"®; we were fortunate to have Jack Canfield and Mark Victor Hansen as two of our motivational speakers. It was while they were still trying to get any publisher to take them seriously!

As part of Peak Performance Institute, Larry and I co-developed "Be Menopositive" which was a public information seminar on Menopause, which has now expanded to Andropause also. My half of the lecture covers the self-worth and coping issues around aging. In 1999, I received a "Woman of the Year Award" from the Toronto Star for my work with women.

Under the Peak organization, we have participated in several clinical trials on women's health issues, most notably, a number on low libido. I have really enjoyed over the years my relationship with the ladies who enrolled in these trials.

I have always wanted to make a difference in people's lives. It was a "Chandler" family tradition mandate. Both my parents, Evelyn and George, were heavily involved in church, senior's organizations and The Legion. They held positions of leadership and influence post-retirement, until they passed away. Dad was instrumental in getting the Remembrance Day poster and essay programs at the Ontario schools. Mom became an advocate for seniors, traveling all over Ontario, advising senior's groups.

I have been so fortunate to have had very strong female role models in my life.

My grandmother, Alice Maud Mary Dunningham, came to Canada as a young woman from England to marry a young man she barely knew who had come to Canada to settle. His home was meager. They had a daughter, Evelyn, my mother. When she was two, her father was killed on the front during World War I. In those days there was no army widows' pension. Making a life for her and her daughter, my grandmother raised chickens. She took them by train to a wealthy clientele that she had developed in the city.

She had heard about a new company called Purina, who made feed. Giving her chickens the brand new feed grew them bigger and better. Her poultry was in huge demand. Purina heard about this amazing woman and asked her to go on the road for them as a salesperson. This was in the 1920s. She declined because she had a young daughter to bring-up at that time. She was women's-libber long before it became popular. She was strong and resourceful and saw to it that her daughter was educated, exposed to culture, and had a good life. She provided a home for her daughter and her two young boys when our father served in the army in WW2. We laugh, because she always referred to any couple using the female's name, but only an initial for the male. It was "Joan and L. Kim and the boy."

My grandmother was the one from whom I inherited the "crafty" genes. She was a prolific knitter, creating vests for third-world children, even after losing her sight completely. She sang in the church choir, was an active member of the Women's Institute and the community. She was a widow her whole adult life. She passed away in her 102nd year. I asked her why she never married, and she said, "I didn't have much, but what I did was mine. You don't know how somebody is going to change things." She was interested in the world, and stayed mentally sharp until the end. She was always dressed beautifully. After the early years, she never wore black but instead, always bright colours. When her vision failed, I bought her cheerful clothes for her, knowing her style. She was always very elegant. This was probably a result of having been bought up by a Victorian-era grandmother. She always wore earrings. I always wear earrings. I do not feel comfortable any other way. She was always there for me. I hope someday I will be able to be a grandmother like her.

Most influential has been my mother, Evelyn. She was one of those people who dealt with every setback in her life with a positive attitude of acceptance and fortitude. She dug in, to do whatever was necessary to carry on. Her motto was always, "Keep on Keeping On" and "This Too Will Pass." She had the strength of character and a determination that was awesome.

Having trained as a Registered Nurse, (though she wanted to be a doctor), she was a stay at home Mom for my brothers and me. It was always so comforting to smell dinner cooking when I came home from school. I wanted my children to feel that way too. She was always ready and willing to listen to my concerns, to chip in and help research a topic of interest. She would drop everything she was doing to get out the resource books. A brilliant student, she was advanced several grades in school. With the local doctor being many miles away in another village, she became the village medical person, even when she was not officially practicing anymore.

After I was married, she decided she wanted to go back to work. She felt nursing was impossible, as she was no longer up to date. At 55, she was hired as a one-woman baby-sitting coordinator for a day-care program for children with cerebral palsy. It was a respite for parents of seriously physically handicapped children, in a little room in the basement of the local hospital. Her salary was paid by the Rotary Club.

To her, it was not right. She felt these children should also be getting therapy, both speech and physical, while they were there. Fierce and determined, she approached the Health Ministry, and physical therapists were hired. Still, this was not good enough; she strongly felt these children should be educated, and so even more determinedly, she approached the Ministry of Education and the local school board. They were granted half of a school that was sitting empty, and they became officially The Landsdown Children`s Center, named after the school they were now occupying. They had many more physically challenged children, and now they had teachers. She was a true coordinator now with staff, budgets, and volunteers. At her funeral, at 92, those who had been part of the school, parents and kids alike, all said the same thing. "Mrs. Chandler always had time for you; If you were a child who had to be toileted and needed help, or if you were a parent who needed a shoulder to cry on, or if you were a staff that needed support (or a place to go for Christmas), she was there and without judgment."

The school became the prototype for the treatment and education of physically challenged children. On top of her running this school, she was asked by the Ministry of Health to go to other communities to talk about their success. They outgrew their adopted school. A beautiful large modern facility was designed and built for them. Having turned 65, and after 10 years of brilliant service, Mrs. Chandler was required to retire. They had to hire 3 people to do her job. There is a large beautiful picture of Mrs. Chandler, Coordinator and Founder, on the foyer wall of the new Landsdown Children`s Center.

She learned to drive when she was in her 60s. When my father died, she was 75, and she became the delighted designated driver for her friends, traveling everywhere. She had always had an incredible sense of direction, the original GPS, and so she never needed a navigator. A serious complication of her eye health forced to give up her cherished licence at 80.

Painful arthritis and mobility problems meant giving up her house for a nursing home. It was just one more thing she handled with grace and acceptance. She was consistently the good patient listener, and often the staff could be found in her room gaining comfort and support.

She was always reading until she fell victim to the family curse, macular degeneration and glaucoma. I arranged audio books for her. She had a wonderful mind until she suffered some sort of cerebral accident just before her 92nd birthday. She passed away a couple of weeks later.

She truly was an amazing person in every respect. She was smart, kind, generous and loving.

I have been blessed with so many inspirational people in my story.

You may contact Joan at joan@peak-performance-institute.com

Joan Chandler Komer's CV is available on the website www.drkomer.com

ABOUT THE CONTRIBUTORS

William S. Cook, Jr., M.D.

Dr. Bill Cook is a board-certified psychiatrist, with a private outpatient practice in Jackson, Mississippi. In addition to his primary practice in Jackson, Dr. Cook serves as a geriatric psychiatrist in a hospital and runs a suboxone clinic at Miss-Lou Addiction Clinic in Natchez. He is the co-author of, "Your Third Act – A Guide To A Great Retirement" for those approaching, and in retirement. Coming in 2017, "Discovering Your Oasis – Hope for Weary Caregivers" is his next book, for those facing stresses in the helping professions.

www.Silverwoods-Publishing.com
@DrWilliamSCook

Blair Lamb, MD, FCFP

Dr. Lamb is a physician who practices in the field of rehabilitative pain medicine whereby we use non-interventional and interventional modalities to facilitate in the rehabilitation of acute and chronic, regional and global, pain and injuries. Dr. Lamb has patents in spinal rehabilitation. Burlington, Ontario, Canada.

www.drlamb.com
www.spinalsolutions.ca
@DrBlairLambMD

Andrew Marr
Warrior Angels Foundation
Co-founder and CEO

Andrew Marr is a Husband, Father, Entrepreneur, former Special Forces Green Beret, founder of Warrior Angels Foundation

www.warriorangelsfoundation.org
@WAFTBI

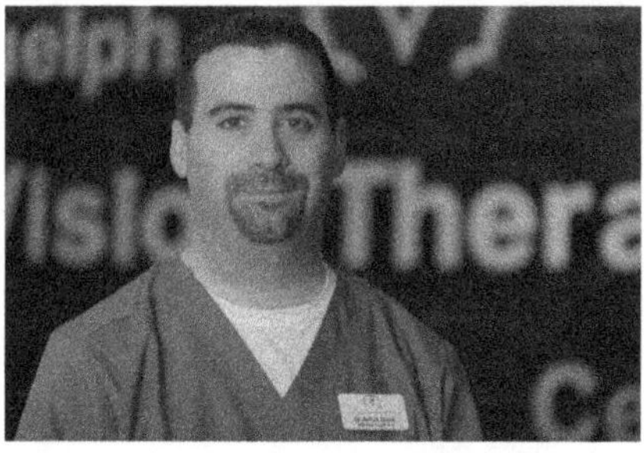

Dr. Patrick Quaid, Neuro-Optometrist, FCOVD, PhD

Appointed Consultant Optometrist, University of Toronto (David L. MacIntosh Sports Medicine Clinic) Adjunct Faculty (Assistant Professor), University of Waterloo School of Optometry & Vision Science Chief

Optometrist & CEO, Guelph Rehabilitative Vision Therapy Centre, Ontario, Canada

If anyone wishes to see us in Guelph (Ontario, Canada) for a consultation simply go do www.gvtc.ca (Guelph Vision Therapy Center) and fill in the questionnaires on-line and submit with your email address and we will contact you ASAP. You can also go to www.covd.org and search across the USA and Canada (be sure to tick the FCOVD or "Board Certified" box to ensure you are dealing with board certified eye doctors in this area as the assessment is quite different to a routine eye examination). There is also a pdf downloadable if you have a child with a reading based learning issue or behavioral (i.e. ADHD-type symptoms) as these issues are also quite commonly found in children with learning issues. If you go to www.gvtc.ca you will also see a blue tab on the right side of the webpage, which you can click on. When you fill in the short form, a free link to a lecture from myself will be emailed to you for your viewing to better educate the public on the impacts of abnormal eye teaming on academics and also in concussion arenas.

SOCIAL MEDIA

THE KOMER BRAIN SCIENCE INSTITUTE

www.komer-brain-science-institute.com
@KTBIDrKomer

DR. LARRY KOMER

www.drkomer.com
@drkomer

JOAN CHANDLER KOMER

www.peak-performance-institute.com
@joankomer

KOMER WOMEN'S HEALTH CENTER

www.komer-womens-health.com
@womenshealthdoc

MASTERS MEN'S CLINIC

www.mastersmensclinic.com
@MastersMensDr

PEAK PERFORMANCE INSTITUTE INC.

Developing the Mind to Improve Performance
www.peak-performance-institute.com

www.ingramcontent.com/pod-product-compliance
Lightning Source LLC
Chambersburg PA
CBHW031404180326
41458CB00043B/6614/J

* 9 7 8 0 9 9 5 2 5 0 1 2 3 *